Safety 2.1

The Safety Envelope

Paul Reyneke

Adapto Publications

Published by Adapto Publications
Website: www.adapto.co.nz

A catalogue record for this book is available from the National Library of New Zealand.

ISBN 978-0-473-72153-4 (paperback)
ISBN 978-0-473-72154-1 (EPUB)

We are what we repeatedly do.
Excellence, then, is not an act but a habit.

—Aristotle (348–322 BCE)

Contents

Part 1
Theoretical Framework

1. INTRODUCTION TO THE SAFETY 2.1 FRAMEWORK   3

2. SAFETY 1 VS SAFETY 2   6
   Model 1   6
   Model 2   7
   Theory and Practice   10

3. COMPLEX ADAPTIVE SYSTEMS   12

4. SAFETY 2.1 THEORETICAL FRAMEWORK   17
   Theory and Practice   22

Part 2
Risk Management

5. INTRODUCTION TO RISK MANAGEMENT   27
   Risk Matrix   27
   Assessing Risk Through the Lens of the 'Reasonable
   Person'   30

6. RISK TOLERANCE   31
   Specific Definition of Tolerable Risk   32
   Maximum Tolerable Risk   36

7. WORKPLACE HAZARDS   38
   What are Workplace Hazards?   38
   Understanding Hazard Severity Assessment   44
   Controlling Hazards   45

8. WORKPLACE RISKS   48
   The Role of Risk in the Risk Matrix   48
   Critical Factors Influencing Risk Assessments   49
   Likelihood Ratings and Factor Definitions   51
   Risk Controls   53
   Word of Caution   54

9. SPIDER DIAGRAM 55
   Spider Matrix Template 56
   Likelihood Reduction 57
   Navigating Control Effectiveness 58

Part 3
The Safety Envelope

10. SAFETY ENVELOPE DEFINED 63

11. OPERATING INSIDE THE SAFETY ENVELOPE 67
   Working Environment and Tools 68
   Skills and Knowledge 69
   Safety Culture 69
   Mental Ability 70

12. WORK ENVIRONMENT AND TOOLS 71
   Work Environment 71
   Tools and Equipment 74

13. VOCATIONAL TRAINING 75

14. CULTURAL BEHAVIOUR 78
   Behavioural Functionality 80
   Understanding Automaticity 80
   Role of Cognitive Processing 82
   Collective Behavioural Patterns 82
   Operational Versus Leadership Roles 83
   Pattern Attributes 84
   Limitations of Cognitive Thinking in Altering Patterns ... 86
   Culture Change Approaches and Challenges 87
   Safety Management Application 91

15. MENTAL WELLNESS 96
   Mental Wellness at Work Model 97
   A. Organisational Factors 98
   B. Environmental Factors 100
   C. Personal Factors 101
   Implementation 102
   Measurement of Wellness 105

Part 4
Summary and Concluding Thoughts

16. THEN THE PENNY DROPS   109

    The Penny Drops   111

    Extending Beyond Safety Management   113

    Bibliography   115

    About the Author   119

Part 1
Theoretical Framework

Chapter 1

Introduction to the
Safety 2.1 Framework

Health and safety is striving to establish itself as a recognised profession. Traditional hallmarks of a profession include specialised knowledge and education, ethical standards and a degree of autonomy based on expertise. However, the field of health and safety often struggles to meet these standards, particularly in aligning theoretical knowledge with practical application.

A key challenge is the disconnect between academic research in health and safety and its implementation in practice. This gap is illustrated by the theoretical concept of 'Safety 2', an innovative approach developed by thinkers like Erik Hollnagel, Sidney Dekker and others. While this approach has been enthusiastically received by practitioners, they often struggle to translate its principles into tangible changes in safety practices. This disconnect is eloquently described by Dekker,[*] who notes the tendency for practitioners to seek prescriptive solutions rather than engaging deeply with new

---

[*] Dekker, S. (2018). *I am not a Policy Wonk*. Blog via www. safetydifferently.com.

theoretical frameworks and adapting them to the complexities of real-world situations.

Another factor is that most safety practitioners operate at a technician level. While professionals are, as noted, generally characterised by advanced education, adherence to ethical standards, and a degree of autonomy in decision-making, technicians focus more on practical, technical skills. This dichotomy is not so much an issue among entry-level safety advisors, but it often leads to a de facto focus on technical compliance over comprehensive safety strategies at senior management level.

The consequences of this divide are particularly noticeable in industries like construction. For instance, in New Zealand, major construction firms demand detailed 'Site Specific Safety Plans' (SSSPs) from subcontractors. These plans, often voluminous and procedural, are mistakenly believed to be a legal requirement. This underscores a broader misunderstanding: the notion that safety can be assured primarily through procedural rigour, without a nuanced understanding of the shared responsibilities and the dynamic nature of safety.

This notion has far-reaching implications for overall safety management. Safety practitioners are often more interested in the newest off-the-shelf tools, and using legalistic arguments, rather than progressing past the old-fashioned safety practices.

Dekker et al.* state that this traditional framework is based on linear cause and effect. Practitioners and line managers focus on what went wrong, leading to an injury, and then work backwards in a

---

* Dekker, S., Cilliers, P., & Hofmeyr, J. H. (2011). The complexity of failure: Implications of complexity theory for safety investigations. *Safety Science* 49(6): 939–945.

straight line to identify the root cause. They search for what Dekker calls the "eureka part" that ultimately failed, following a reductionist method.

Dekker continues to contrast this with Safety 2-thinking: "Analytic reduction cannot tell how a number of different things and processes act together when exposed to a number of different influences at the same time. This is complexity, a characteristic of a system. Complex behaviour arises because of the interaction between the components of a system. It asks us to focus not on individual components but on their relationships. The properties of the system emerge as a result of these interactions; they are not contained within individual components."

This book aims to bridge this crucial gap in the field of safety. While not an exhaustive exploration of every aspect of safety, it seeks to highlight the important connection between theoretical insights and practical applications. Notably, it emphasises the synergy between the profound contributions of theorists and the real-world challenges faced by practitioners. Effective feedback mechanisms are essential for theorists, allowing them to refine theories that may falter over time. Conversely, practitioners should actively incorporate the latest theoretical advancements to enhance safety practices.

As someone who proudly identifies as a practitioner, yet has had the privilege of collaborating with theorists, the author primarily addresses fellow practitioners. The purpose is to provoke thought, challenge existing safety practices and contribute to the advancement of the safety profession as a whole.

Chapter 2
Safety 1 vs Safety 2

There are currently two very different approaches to safety management. It is not always clear if this is the result of a deliberate choice or not; however, the approaches are very different. Andrew Hale and David Borys* articulate the two approaches very clearly and the following is an almost verbatim extract from their report to the Institution of Occupational Safety and Health (IOSH) Research Committee.

They refer to the different approaches as "Model 1" and "Model 2" and summarise the models as follows:

Model 1

This model is rooted in scientific management. It is rationalist and prescriptive in its approach, and it sees rules as the embodiment of the single best way to carry out activities covering all (most) known contingencies. Rules are devised by experts to guard against the

* Hale, A., & Borys, D. (2013). Working to rule or working safely? Part 2: The management of safety rules and procedures. *Safety Science* 55: 222–231.

errors and mistakes of fallible humans at the sharp end (the operators), who are more limited than the experts in working out the best way to do things.

Rules are essentially created top-down, should be decided in advance, and be based on task and risk analyses. Once devised, they are 'carved in stone', communicated to and imposed on the workforce by management. Violations (intentional deviations) and errors (unintentional deviations) are seen as essentially negative actions that should be countered and suppressed, as a last resort through punishment.

Rules are to be documented in manuals or databases and consist mainly of abstract must-statements. Language is formal and precise to avoid ambiguity. It is common to include extensive sections defining terminology and referencing sources. It is then made available to the workforce in the form of instructions, incorporated in training and signed for by operators to signify their intent to comply.

Local managers and supervisors are expected to take their enforcement roles seriously and non-enforcement of the rules may also be subject to sanctions.

Model 2

This model sees rules as patterns of behaviour, socially constructed, emerging from the experience of those carrying them out. They are characterised as local and situated in the specific activity, in contrast to the written rules, which are seen as being at a generic level, necessarily abstracted from the detailed situation to be able to generalise them across essentially disparate local situations.

The Model 2 view of rules is essentially bottom-up and dynamic. It recognises that written rules, except for a few 'golden rules' typically prescribed by law or industry best practice, require a process of translation and adaptation before application to any given, specific situation. This implies that written rules should not be at the detailed, action level but, at most, at the process level.

The real experts in this conceptualisation are the operators, whose ability to conduct and navigate this dynamic process of negotiation and construction of rules is seen as an essential part of their skill and identity – and they should be trained and developed to fulfil this expectation. A likely response to attempts to impose rules from outside of this operational group is resistance. While informal and group rules are seen as violations by those on the outside, they are seen as skilled adaptations by operators.

Rules are seen in model 2 as support and guidance for the expert. However, except for the few 'golden rules' mentioned above, they are not something requiring strict compliance and no substitute for competence, unless the operator is a novice/not yet fully skilled.

'Violation' is therefore seen as essential in specific cases where the rule does not match the reality and is part of applying the mature operator's tacit knowledge in the process of carrying out work, not a form of defiant behaviour.

Enforcing the rule without allowing for adaptation to the local reality is seen as punitive, demeaning and destroying trust, and it is scoffed at by experienced operators, at least in private.

Based on this analysis, Hale and Borys postulate the following 'principles':

- Rules, documented or not, are necessary, but they are not the only or even the preferred solutions to ensure control.

Design, competence and social control are also valid alternatives to the written word.

- Rules influencing safety should be combined with rules on quality, productivity, sustainability and so on to reduce the size of the rulebook.
- Rule violation is a signal that all is not well with the rule system and that punishment, or discipline, may be inappropriate. Revisiting the possible disconnect between the rule and the task reality may shed more light on the violations. Compliance is important, but understanding non-compliance is more important.
- The participation of competent and motivated rule users in making and improving rules is essential. Experienced operators are the experts in rule use in real-life situations.

The model is hierarchical:

1. At the top are goals specifying only the outcomes.
2. Next come process rules concerning how to translate the goals into acceptable processes.
3. Finally, action rules (exact behaviour prescriptions) at the bottom restrict the freedom of choice of rule users.

Rules at the higher end of the hierarchy (goals and process rules) place more trust in the rule users to translate them to each situation. On the other hand, rules at the lower end of the hierarchy (action rules) lead to more exceptions and 'violations' in different situations.

Rulemaking is a balance between these two concerns.

- In most situations, there is more than one way to carry out a task safely. What matters is keeping within a 'safe envelope' of behaviour. For competent rule users, rules can be phrased simply as guidance, unless the operator knows a way of behaving that is at least as safe and achieves the same result. This could reduce the number of 'golden rules' which must be followed exactly.

- People experienced in using their judgement to decide the appropriate behaviour in their normal work are more likely to be able to work out what to do when faced with unexpected and unpredicted situations.
- In activities where people have to work together, there is a stronger case for a more central formulation of rules, so that everyone can predict what teammates will do. This can be done at an organisational level, such as determining traffic systems, shared site rules and emergency procedures, or it can be agreed by the social group.
- Rulemaking can never be abdicated; it remains an organisational process. Even if rules are produced in a bottom-up manner, managers must ensure rules do not drift to the boundaries of the safe envelope.
- All rules have a lifespan and become potentially outdated over time. All rules must regularly be reviewed to avoid them becoming disconnected from the reality of task demands.

Theory and Practice

The Safety 2 theory presents an intuitive yet sophisticated approach to workplace safety. Its core premise is often readily acknowledged and accepted by many in the safety management industry. However, the real challenge in leveraging the Safety 2 theory lies not only in its conceptual acceptance but also in the intricacies of its application in real-world scenarios. It is imperative to gain a comprehensive and nuanced understanding of its fundamental principles when implementing this theory.

A pivotal aspect is its reliance on the concept of complex systems and, more specifically, complex adaptive systems. This concept is crucial because it acknowledges that organisations are not static; they are dynamic and continually evolving. The following chapter, while steering clear of an overly academic or technical narrative, aims to demystify these core principles. By doing so, it provides a fundamental understanding that is indispensable for those looking to apply the theory in a practical, impactful manner.

This application is in essence the practical application of Safety 2, hence the extension to Safety 2.1.

The next chapter seeks to start equipping practitioners with the necessary insights not only to understand the complex nature of their organisational systems but also to navigate and influence these systems towards a more proactive safety culture. This understanding is key to unlocking the full potential of the Safety 2 theory, turning it from a conceptual framework into a tangible, effective tool for improving workplace safety.

Chapter 3
Complex Adaptive Systems

The Safety 2.1 approach in this book is based on complex adaptive systems theory, focusing on how safety elements interact dynamically and adapt in real-world situations.

Richard Pascale,[*] a pioneer in the study of complex adaptive systems, has highlighted a paradigm shift in our comprehension of organisational operations over the previous two to three decades.

During the 1980s and 1990s, organisational focus was predominantly on performance enhancement as evidenced by methodologies such as total quality improvement, kaizen, just-in-time, and re-engineering. However, Pascale proposed that our understanding of how business works has moved beyond these concepts. He and others, like Saskia Harkema,[†] critiqued this approach, noting its reliance on linear cause-and-effect assumptions. It assumes that if something is not performing as it

[*] Pascale, R. T. (1999). Surfing the edge of chaos. *Sloan Management Review* 40(3): 83-94.

[†] Harkema, S. (2003). A complex adaptive perspective on learning within innovation projects. *The Learning Organisation* 10(6): 340-346.

should, all that is required is to discover what went wrong, correct it and success will follow.

Such linearity is not a feature of complex systems, for example health and safety systems. Health and safety is inherently subjective, and outcomes stem from multifaceted interactions among various elements – people, the environment, social dynamics and others. These interactions are unique and non-repetitive, making it impossible to comprehend the system fully by analysing individual components in isolation.

Complex adaptive systems, a distinct category within complex systems, is distinguished by the adaptive and learning capabilities of its components. Over time, complex adaptive systems evolve specific rules governing agent actions and their interconnections, resulting in diverse aggregate outcomes. These outcomes, or 'emergent properties', are the manifestations of the interplay of agents' behaviours.

Key Characteristics of Complex Adaptive Systems

1. **Emergence**: A defining characteristic of complex adaptive systems is that they exhibit properties and behaviours that go beyond their individual parts, a phenomenon known as emergence. This means that the collective characteristics of the system cannot be predicted merely by analysing its components. For instance, the complex behaviour in a flock of birds – their coordinated, fluid movements – is not something you can understand by examining a single bird. Emergence reflects how interactions at a micro level can lead to unexpected macro-level behaviours, revealing the intricate interconnectedness of the system's elements.

2. **Self-Organisation**: Complex adaptive systems have the remarkable ability to organise themselves autonomously,

without any centralised control or external guidance. This self-organisation is often a response to changes in the environment and can result in the development of new structures or behaviour patterns. Take ant colonies as an example: these colonies exhibit sophisticated structures and collective behaviours that arise from the simple rules followed by individual ants. This characteristic highlights the system's inherent capacity to adapt and restructure itself.

3. **Adaptation and Evolution**: A key trait of complex adaptive systems is their ability to adapt and evolve in response to environmental interactions. This evolutionary process enables the system to modify and continue existing under diverse conditions. In the field of economics, markets are a good illustration of this feature. Markets continuously adapt to new information, technologies and regulations, showing an evolutionary progression over time.

4. **Non-linearity**: Relationships within complex adaptive systems are typically non-linear, meaning that the system's responses to changes can be disproportionate and unpredictable. This non-linearity is often exemplified by the "butterfly effect", where even minor changes can lead to significant, unforeseen consequences. This aspect underlines the complexity of predicting the behaviour of these systems, as small inputs or modifications can result in either negligible or substantial impacts.

5. **Feedback Loops**: Complex adaptive systems are characterised by the presence of both positive and negative feedback loops. These loops are essential for the system's stability and adaptability, as they enable the system to adjust its behaviour based on the outcomes of these feedback mechanisms. In social systems, for example, public opinion can serve as a feedback loop, significantly

influencing political decisions and societal norms. These loops play a critical role in maintaining a dynamic equilibrium within the system.

6. **Edge of Chaos**: Complex adaptive systems often operate at the edge of chaos, a state balanced between order and disorder. This critical juncture is where the system exhibits its highest level of adaptability and capability for complex behaviour. Being at the edge of chaos is akin to being in a zone of optimal tension and creativity, where the potential for innovation and the emergence of new patterns is greatest. It is this delicate balance that enables the system to evolve and thrive in a constantly changing environment.

These characteristics collectively contribute to the dynamic and intricate nature of complex adaptive systems, making them a fascinating subject for study and analysis in the safety management field.

Perceiving safety management as a complex adaptive system offers a revolutionary perspective. It acknowledges that safety is not a static target but a dynamic, ever-evolving process. This process continuously adapts and transforms with each interaction and decision within the organisational ecosystem. Unlike traditional linear models that emphasise direct cause-and-effect relationships, the complex adaptive systems approach embraces a holistic view. It recognises safety as an emergent property, born from the intricate web of interactions and relationships within the system.

This paradigm shift necessitates adopting a systemic approach, where the interdependence of various components is acknowledged. In such a framework, changes in one part of the system can ripple through and impact the entire organisation. It acknowledges that learning and adaptation occur at a systemic level, transcending individual efforts. This perspective fosters a culture where learning is shared and integrated into the organisational fabric, to enhance the collective knowledge base.

Moreover, the self-organising principle of complex adaptive systems empowers employees at all levels. It encourages autonomy and decision-making at the ground level, fostering a sense of ownership and responsibility towards safety. This decentralised approach allows for more agile responses to safety challenges, as employees are more attuned to the nuances of their immediate environment and can take swift, informed actions.

In such a system, the emphasis is on the collective over the individual. It is an acknowledgment that the whole organisation is more than just a sum of its parts. The interactions, shared knowledge and collaborative efforts contribute to a more resilient and adaptive safety culture. This approach to safety management aligns with the dynamic and ever-changing nature of organisations, ensuring that safety protocols and practices are not only responsive but also proactive in identifying and mitigating risks.

The integration of complex adaptive systems theory in safety management is thus a transformative step, moving away from rigid, prescriptive models to a more fluid, adaptive approach. It aligns with the realities of modern organisational dynamics and paves the way for a more holistic and effective safety culture.

The evolution towards Safety 2.1 epitomises this journey. It builds upon a robust theoretical foundation while primarily concentrating on the practical implications and real-world applications. This shift marks a pivotal move from theoretical understanding to actionable strategies.

Chapter 4
Safety 2.1 Theoretical Framework

The conceptual model presented here, termed Safety 2.1, evolves from the foundational principles of Safety 2 as proposed by Hollnagel et al.,[*] Dekker,[†] Conklin[‡] and many others. Hale and Borys are also in this camp. This iteration – 2.1 – adds to typical Safety 2 by uniquely integrating theory with practice, addressing a critical gap observed in earlier models.

Safety 2, and therefore Safety 2.1, in contrast to the traditional Safety 1 approach, places a significant emphasis on allowing workers to control how they will do work, rather than prescribing every detail. It recognises the complexity of work and allows for collaborative work methods.

---

[*] Hollnagel, E., Wears, R. L., & Braithwaite, J. (2015). *From Safety-I to Safety-II: a white paper. The resilient health care net*: published simultaneously by the University of Southern Denmark, University of Florida, USA, and Macquarie University, Australia.
[†] Dekker, S. (2014). *Safety differently: Human factors for a new era*. CRC Press.
[‡] Conklin, T. (2019). *Pre-accident investigations: An introduction to organisational safety*. CRC Press.

Safety 2.1 specifically aims to address safety practice and can be contrasted with traditional safety practices – Safety 1 – as follows:

Behaviourism vs Complex Adaptive Systems: Safety 1 operates on a behaviourist model, emphasising shaping actions through reward and punishment and reinforcing specific behaviours for safety compliance. Safety 2.1, on the other hand, considers management as a complex, socially constructed system. It fosters change not just through direct interventions but by leveraging its inherent pattern-forming abilities, embracing the dynamic interactions and emergent properties of organisational networks.

Rule-Makers vs Change Agents: In the Safety 1 paradigm, safety practitioners are predominantly seen as makers and enforcers of rules, tasked with ensuring adherence to established safety protocols. Safety 2.1 reconceptualises their role as change agents who facilitate adaptation and growth. They are seen as catalysts in a dynamic environment, empowering workers and encouraging innovative solutions to safety challenges.

Limited Best Practices vs Multiple Approaches: Safety 1 prescribes a limited number of 'best' methods for ensuring safe work, often leading to a one-size-fits-all approach. In contrast, Safety 2.1 recognises the uniqueness of each situation and promotes flexibility in approach. It acknowledges a broader spectrum of effective strategies, tailored to specific contexts and challenges.

Hierarchical vs Sapiential Authority: Safety 1 is characterised by a hierarchical approach, emphasising top-down authority and decision-making. Safety 2.1, however, values the competence and wisdom (sapience) of frontline technicians. It appreciates the depth of their experience and insights, encouraging a more collaborative and inclusive approach to safety management.

Objective Risk Assessment vs Unpredictable Risks: In Safety 1, risk is perceived as objectively measurable and quantifiable, suggesting a predictable and controllable environment. Conversely, Safety 2.1 recognises the unpredictable and complex nature of most risks, especially in dynamic and rapidly evolving contexts. It promotes a more nuanced understanding of risk as a multifaceted and often unpredictable phenomenon.

Expert-Created Rules vs Technician-Decided Rules: In the Safety 1 model, rules and protocols are typically defined by experts who may be removed from the day-to-day realities of the operational environment. Safety 2.1, however, advocates for a more bottom-up approach, where technicians with hands-on experience play a crucial role in crafting the guidelines. This approach values the practical wisdom and contextual knowledge of those directly involved in the work.

Increasing Rigidity in Rules vs Enhancing Frontline Competency: Safety 1 often responds to incidents or failures by implementing more rigid rules, potentially leading to an over-regulated environment. Safety 2.1, in contrast, focuses on enhancing the competency and decision-making skills of frontline workers. It emphasises the development of their ability to respond to unexpected situations, thereby fostering a more adaptive and resilient safety culture.

Valuing Consistency vs Emphasising Resilience: In Safety 1, consistency is key, often pursued through rigorous document control and strict adherence to protocols. Safety 2.1, however, places greater emphasis on resilience and adaptability in changing situations. It recognises the importance of being able to adjust and respond effectively to unforeseen challenges, rather than simply maintaining a static set of practices.

Static Rules vs Adaptive Guidelines: Safety 1 is characterised by setting inflexible rules that are often applied uniformly across various situations. Safety 2.1, on the other hand, prioritises adaptability, setting only a few non-negotiable principles and allowing for flexibility within safe operational boundaries. This approach acknowledges the need for situational judgement and contextual decision-making.

Cognitive Thinking vs Automatic Mind Patterns: Safety 1 approaches work through a lens of measurable, cognitive thinking, where actions are subject to deliberate reasoning and alteration. Safety 2.1 recognises that much of the work is conducted through ingrained, automatic mind patterns that operate beyond conscious thought. It appreciates the role of intuitive and experiential knowledge in guiding actions, especially under pressure or in complex situations.

Rule Enforcement vs Safe Envelope Management: In Safety 1, local management is primarily viewed as enforcers of rules, focusing on compliance and adherence to predefined standards. Safety 2.1, in contrast, envisions them as guardians of a 'safe envelope'. Within this envelope, technicians have the freedom to operate, make decisions and innovate, provided they maintain overall safety. This approach allows for greater flexibility and responsiveness to real-time situations.

Compliance vs Social Enforcement: Safety 1 is centred around achieving compliance from frontline staff, often through formal mechanisms and, if necessary, enforcement actions. In Safety 2.1, safety maintenance is more socially driven, embedded in group norms and behaviours, and guided by knowledgeable peer-group leaders. It relies on the collective responsibility and shared values of the workforce to uphold safety standards.

Patronising Dependence vs Interdependence: Safety 1 often adopts a patronising approach, with safety experts assuming a top-down responsibility for subordinates' actions. Conversely, Safety 2.1 encourages a culture of interdependence, where the collective wisdom and collaboration of all employees contribute to system improvement. This model fosters a sense of shared responsibility and mutual respect among all levels of the organisation.

External Rewards vs Self-Actualisation: Safety 1 typically motivates through external rewards, such as job security and financial incentives. Safety 2.1, however, shifts the focus to intrinsic rewards linked to the self-actualisation needs of frontline staff. It values personal growth, job satisfaction and a sense of accomplishment as key drivers for maintaining safety standards.

Eliminating Errors vs Valuing Adaptations: The aim of Safety 1 is predominantly to eliminate violations and errors, seeking to create a fault-free work environment. In contrast, Safety 2.1 values interpretations, adaptations and innovations by the workforce as higher-order skills. It recognises that in complex systems, adaptability and creative problem-solving are essential for safety and efficiency.

Avoiding Harm vs Consistent Excellence: Safety 1 focuses primarily on avoiding harm, epitomised by the goal of 'zero harm'. Safety 2.1, on the other hand, concentrates on consistently achieving positive outcomes. It seeks not simply to prevent accidents but to create conditions where excellence in safety is the norm.

Mistrust vs Trust: In Safety 1, there is often an implicit mistrust towards frontline staff, who are sometimes perceived as potential sources of unreliability. Safety 2.1, in contrast, builds a culture of trust as a cornerstone of effective safety management. It encourages open, bi-directional communication and values the insights and experiences of all employees.

Objective Measurement vs Social Construct Measurement: Safety 1 relies heavily on objective metrics, such as failure counts, to gauge safety performance. Safety 2.1, however, views measurement as a social construct, evaluating safety through collective judgement and shared understanding. It recognises that safety can be a subjective experience, influenced by cultural and contextual factors.

Audits as Goals vs Audits as Tools: In the Safety 1 approach, audits are seen as imperative, with passing them being an end goal in itself. Safety 2.1, meanwhile, uses audits as tools for continuous improvement. Audits are not just for compliance but are leveraged to ensure behaviours and practices remain within the safety envelope, providing insights for ongoing adaptation and enhancement.

Documentation Focus vs Intergenerational Learning: Safety 1 places a heavy emphasis on extensive documentation for knowledge preservation and regulatory compliance. In contrast, Safety 2.1 values more dynamic forms of knowledge transfer, such as collegial learning and mentoring. It emphasises personal interactions and the sharing of experiences and insights across generations, recognising that safety wisdom is often best transmitted through relational and experiential means.

Theory and Practice

Safety 2.1 is deeply rooted in a variety of theoretical frameworks, making it inherently complex and dynamic. It is characterised by its adaptability, flexibility and a deliberate move away from prescriptive, one-size-fits-all methods. This paradigm shift represents a significant departure from traditional safety methodologies, embracing a more holistic view of organisational safety. Implementing Safety 2.1 effectively requires a continuous and thoughtful reference to its theoretical underpinnings. These principles, which have been mentioned previously, form the backbone of this approach. Neglecting to apply these theoretical concepts consistently risks regression to the more rigid, conventional Safety 1 methodologies. Safety 2.1 is designed to transcend these traditional approaches, offering a more nuanced and responsive way of managing safety that aligns with the complexities of modern organisational environments.

Safety 2.1 strives to add to the Safety 2 framework by specifically focusing on the operationalisation of the key constructs, bridging the gap between these theoretical foundations and their practical application in the field of safety management. While the focus in the following chapters will be on a pivotal aspect of safety management systems – the management of hazards and risks – it is crucial to emphasise that this text is not intended to serve as a step-by-step manual for executing safety-management strategies. Instead, it aims to provide a conceptual framework of guiding principles that can inform and shape practical safety-management approaches. This perspective acknowledges the varied and often unpredictable nature of organisational environments, where rigid procedures may not always be applicable or effective. By understanding and applying the core concepts of Safety 2.1, safety practitioners and organisational leaders can develop more adaptable, resilient and contextually relevant safety strategies. These strategies, while guided by theory, must be tailored to the unique characteristics and needs of each organisation, allowing for a more organic and effective integration of safety into the fabric of organisational life.

This exploration will not only highlight the implementation processes but also the challenges and learning opportunities encountered in implementing Safety 2.1. This aims to equip readers with the insights and tools necessary to navigate the complexities of modern safety management, fostering an environment where safety is not just a compliance requirement but an integral part of organisational culture and performance.

Part 2
Risk Management

Chapter 5
Introduction to Risk Management

Managing risks is a key part of keeping people safe at work. It is about finding, understanding and controlling hazards that could cause accidents or harm. The main aim is to plan safety measures in a way that prevents accidents and reduces harm. This means looking at all aspects of safety management and working to prevent, or at least mitigate, harm.

Risk Matrix

A tool commonly used to do this is a five-by-five risk matrix. It helps to classify risks based on two factors: how severe potential harm could be (ranging from "insignificant" to "catastrophic") and how likely it is that somebody could be harmed (ranging from "highly unlikely" to "highly likely"). It serves to translate abstract safety principles into concrete, actionable steps.

Variable Risk Tolerance	LIKELIHOOD RATING				
	1	2	3	4	5
	EXTREMELY RARE	RARE	POSSIBLE	LIKELY	ALMOST CERTAIN
SEVERITY RATING	Highly unlikely; cannot rule it out	Conceivable but unlikely	Might happen at some time	Likely would happen sometime	Highly likely/ expected to happen
5 CATASTROPHIC Fatality	15	19	22	24	25
4 MAJOR Permanent Incapacity or Life-changing injury	10	14	18	21	23
3 REVERSIBLE Requires time off work	6	9	13	17	20
2 MINOR Medical treatment/ restricted duties	3	5	8	12	16
1 INSIGNIFICANT Only first aid treatment	1	2	4	7	11

Figure 1. Typical 5x5 Risk Matrix.

While the risk matrix is a useful tool, it has several challenges:

Ambiguous Definitions of the Axes: The matrix's scale, especially the likelihood scale, which measures how probable an event is, often lacks clarity and precision. This vagueness hampers the process of making objective assessments. While statistical probabilities are frequently employed in these assessments, their interpretation varies significantly among different users. This variability in interpretation contributes to the ambiguity in assessing the likelihood of risks, making the process less definitive and more subjective.

Subjectivity in Rating Risks: The process of risk assessment is influenced by various subjective factors, such as the individual risk tolerance of those involved in the assessment and the overarching culture of the organisation. These subjective elements can lead to inconsistencies in risk ratings, thereby affecting the reliability and uniformity of the assessments across different scenarios and departments within the same organisation.

Size of Risk Reduction: A common issue in risk assessment is the tendency to overestimate the effectiveness of certain safety controls, particularly those that are less tangible, like administrative controls. This overestimation can result in a skewed perception of the residual risk that remains even after the implementation of these controls. Consequently, this can lead to a false sense of security regarding the safety measures in place and an underestimation of the actual risk that persists.

Misunderstanding the ALARP Principle: The principle of 'As Low as Reasonably Practicable' (ALARP) is an acknowledgement that some level of risk is inevitable and cannot be entirely eliminated. However, this principle is often not represented in risk-assessment matrices, leading to confusion regarding whether additional controls are necessary. The lack of visibility of the ALARP principle in these matrices can result in either excessive or insufficient safety measures, as it becomes challenging to determine the point at which risk reduction has been maximised within reasonable bounds.

Difference between Hazards and Risks: There is often a misunderstanding or confusion between the concepts of hazards and risks, with these terms being used interchangeably. However, they represent fundamentally different elements within the realm of safety and risk management. Hazards refer to potential sources of harm or adverse health effects, while risks are concerned with the likelihood and impact of these hazards manifesting. Misconstruing these terms not only affects the accuracy of communication but also influences the approach to risk management, as strategies for addressing hazards might differ significantly from those for managing risks. This subtle yet significant distinction warrants further exploration and clarification to enhance the effectiveness of risk-management practices.

Assessing Risk Through the Lens of the 'Reasonable Person'

As mentioned before, when assessing risk, it is important to understand that people have different risk tolerance levels regarding how much risk is acceptable. Some people may have an overly pessimistic view of risk and expect the most severe outcomes. It is a very cautious approach, but it can make risks seem bigger than they are. Another group of people may be over-optimistic and expect the best possible outcome. This view might miss some important risks because it is too hopeful.

The third possibility is to find a middle ground. It uses evidence and realistic thinking to establish the most likely level of harm. This method is similar to how a 'reasonable person' would think about risk, and it is usually the best way to assess risks accurately.

It is important to approach the concept of 'middle ground' with caution. This term does not simply imply a halfway point between the best and worst scenarios. Rather, it represents a search for a balanced perspective, combining elements of optimism and pragmatism. It is about identifying the 'realistically probable scenario', which comes from carefully considering all possible options.

Using experts with specialised knowledge of the hazards can help to improve assessments. However, the decision is not simply based on data; it is ultimately a managerial judgement that could improve over time.

Chapter 6
Risk Tolerance

Risk tolerance is not only an individual reality; that is, some people are less risk averse than others. It is also an organisational phenomenon that should be considered during risk-assessment processes.

Risk mitigation often comes at a cost. In fact, there is an ever-present and inevitable conflict between safety controls and production; safety control measures almost always inhibit production. When these controls do not interfere with production, the course of action is straightforward: eliminate or significantly lower the risk. However, this simplicity is not common. The introduction of stricter safety controls often detrimentally impacts production, eliciting resistance from production managers who naturally challenge any reductions in productivity. This dynamic raises the essential question of the optimal balance: what is the organisation's risk tolerance?

'Tolerable safety risk' refers to an acceptable level of risk within a specific context, striking a balance between the necessity of certain activities and the inherent safety hazards they present. This concept involves a thorough evaluation of the potential harm versus the benefits of the operational activity in question, alongside the practicality and effectiveness of risk-mitigation measures. This concept acknowledges that when complete risk elimination is impossible, risks must at least be reduced to a socially and organisationally acceptable level. It must also satisfy legal expectations.

Importantly, embracing tolerable risk is not an acceptance of sub-par safety standards. Rather, it calls for informed decision-making that judiciously weighs risks against benefits in a manner that is ethically sound, economically viable and socially responsible.

Despite the widespread recognition that a level of residual risk will remain after mitigation controls are implemented, a challenge emerges in its application: many organisations fail to explicitly define their threshold of tolerable risk. This often leads to reliance on subjective judgements of what 'feels right', as opposed to striving to reach a predefined level of risk acceptance.

This issue is further complicated by the common use of risk matrices that do not specify the point at which risk remains intolerable. Consequently, the effectiveness of risk-mitigation strategies can be misjudged, sometimes presuming significant risk reduction when, in reality, only marginal safety measures have been implemented, and the risk is not sufficiently reduced.

Specific Definition of Tolerable Risk

Safety 2.1 proposes to explicitly define the level of risk tolerance on the risk matrix, for example:

Risk Tolerance	LIKELIHOOD RATING				
	1	2	3	4	5
	EXTREMELY RARE	RARE	POSSIBLE	LIKELY	ALMOST CERTAIN
SEVERITY RATING	Highly unlikely; cannot rule it out	Conceivable but unlikely	Might happen at some time	Likely would happen sometime	Highly likely/ expected to happen
5 CATASTROPHIC Fatality	15	19	22	24	25
4 MAJOR Permanent Incapacity or Life-changing injury	10	14	18	21	23
3 REVERSIBLE Requires time off work	6	9	13	17	20
2 MINOR Medical treatment/ restricted duties	3	5	8	12	16
1 INSIGNIFICANT Only first aid treatment	1	2	4	7	11

Figure 2. Risk Tolerance.

In practical terms, this means that if a hazard, after the application of mitigation controls, still presents a high risk – an overall risk score of 16 and higher in the above example – the risk remains unacceptable and requires further mitigation. For hazards with a risk score between 11 and 15, each case must be evaluated individually to decide if the risk can be accepted (will be tolerated), or not. This is a 'grey zone' and a matter of professional discretion. The rationale behind these decisions should be thoroughly documented to facilitate future reviews and assessments.

Risk below this threshold (rating of 10 and below) does not require further formal risk controls to be implemented. On the contrary, risk controls, especially introducing safety procedures below this threshold, are often seen by experienced operators as insulting and demeaning.

This does not imply that every organisation will exhibit the same tolerance for risk. The level of risk an organisation is willing to accept could be higher or, conversely, the organisation may exhibit a high degree of risk aversion. As illustrated below, this risk tolerance is also not necessarily a straight line on the matrix.

High Risk Tolerance:

The organisation may decide to accept a higher level of risk, i.e., only truly high risks are controlled, while the rest is left up to the operators to control.

High Risk Tolerance	LIKELIHOOD RATING				
	1	2	3	4	5
	EXTREMELY RARE	RARE	POSSIBLE	LIKELY	ALMOST CERTAIN
SEVERITY RATING	Highly unlikely; cannot rule it out	Conceivable but unlikely	Might happen at some time	Likely would happen sometime	Highly likely/ expected to happen
5 CATASTROPHIC Fatality	15	19	22	24	25
4 MAJOR Permanent Incapacity or Life-changing injury	10	14	18	21	23
3 REVERSIBLE Requires time off work	6	9	13	17	20
2 MINOR Medical treatment/ restricted duties	3	5	8	12	16
1 INSIGNIFICANT Only first aid treatment	1	2	4	7	11

Figure 3. High Risk Tolerance.

Apart from the social and reputational risks associated with this, the legal implications of setting the risk tolerance levels too high must be considered.

High Risk Aversion:

The organisation may on the other hand lower the risk tolerance, i.e., introduce formal controls to lower the risk further.

High Risk Aversion	LIKELIHOOD RATING				
	1	2	3	4	5
	EXTREMELY RARE	RARE	POSSIBLE	LIKELY	ALMOST CERTAIN
SEVERITY RATING	Highly unlikely; cannot rule it out	Conceivable but unlikely	Might happen at some time	Likely would happen sometime	Highly likely/ expected to happen
5 CATASTROPHIC Fatality	15	19	22	24	25
4 MAJOR Permanent Incapacity or Life-changing injury	10	14	18	21	23
3 REVERSIBLE Requires time off work	6	9	13	17	20
2 MINOR Medical treatment/ restricted duties	3	5	8	12	16
1 INSIGNIFICANT Only first aid treatment	1	2	4	7	11

Intolerable Risk

Tolerable Risk

Figure 4. High Risk Aversion.

The consequence of this approach may be that the organisation increases controls, effectively returning to a Safety 1 approach where the organisation extensively prescribes the use of more controls, often administrative controls, and in particular written procedures, to control the risk to a lower level.

Variable Risk Tolerance:

In this case the organisation varies its risk tolerance, whereby they are more risk averse when the hazard presents higher severity risks, and willing to accept more risk for hazards holding lesser severity consequences.

This is a very sensible approach, especially where there is a small number of critical hazards that require more prescribed controls, while the rest can be left to the frontline people to control.

Crucially, it is essential for an organisation to clearly define its acceptable level of risk. This threshold should be rigorously evaluated against what is socially and legally acceptable.

Variable Risk Tolerance		LIKELIHOOD RATING				
		1	2	3	4	5
		EXTREMELY RARE	RARE	POSSIBLE	LIKELY	ALMOST CERTAIN
SEVERITY RATING		Highly unlikely; cannot rule it out	Conceivable but unlikely	Might happen at some time	Likely would happen sometime	Highly likely/ expected to happen
5	CATASTROPHIC Fatality	15	19	22	24	25
4	MAJOR Permanent Incapacity or Life-changing injury	10	14	18	21	23
3	REVERSIBLE Requires time off work	6	9	13	17	20
2	MINOR Medical treatment/ restricted duties	3	5	8	12	16
1	INSIGNIFICANT Only first aid treatment	1	2	4	7	11

Figure 5. Variable Risk Tolerance.

Maximum Tolerable Risk

When assessing the maximum risk an organisation can accept for a particular hazard, the organisation uses the Risk Tolerance Matrix it adopted as a template, as discussed above. As will be discussed in the next chapter, the severity rating – how severe potential injuries could be – is the more enduring feature of the hazard. It therefore provides a more stable foundation for determining the maximum tolerance level; once the severity rating for the hazard is determined, the organisation identifies the corresponding maximum likelihood rating on the matrix, as illustrated below.

In this example, the hazard is assessed to present a potential for causing reversable harm and, as a result, the organisation's acceptable maximum likelihood level would be categorised as Possible. This implies that the organisation is unwilling to tolerate a probability higher than possible for an individual to sustain permanent injuries. Any assessment greater than this possible level would surpass the risk-tolerance threshold, prompting the need for risk-mitigation measures.

Likelihood Conversion	LIKELIHOOD RATING				
	1	2	3	4	5
	EXTREMELY RARE	RARE	POSSIBLE	LIKELY	ALMOST CERTAIN
SEVERITY RATING	Highly unlikely; cannot rule it out	Conceivable but unlikely	Might happen at some time	Likely would happen sometime	Highly likely/ expected to happen
5 CATASTROPHIC Fatality	15	19		24	25
4 MAJOR Permanent Incapacity or Life-changing injury	10	14		21	23
3 REVERSIBLE Requires time off work			13	17	20
2 MINOR Medical treatment/ restricted duties	3	5	8	12	16
1 INSIGNIFICANT Only first aid treatment	1	2	4	7	11

Figure 6. Likelihood Conversion.

The assessments of both the severity of potential harm and the likelihood of this level of harm occurring are the topics of the next three chapters.

Chapter 7
Workplace Hazards

The terms 'hazard' and 'risk' are often used interchangeably in the safety profession, even though they do not describe the same construct. First, it is important to distinguish between a 'hazard' and a 'risk', and then to understand how the differences influence the overall risk-management process.

A hazard denotes any potential source of harm, injury or adverse health effects in a workplace setting, while a risk refers to the probability of the harm occurring.

What are Workplace Hazards?

As mentioned above, a hazard is any potential source of harm but it does not consider the probability of the harm occurring. It can be an object, condition or activity/inactivity that poses a threat of injury or illness. For instance, live electricity is a safety hazard due to its potential to cause electric shocks. However, a description of the hazard – 'live electrical wires' – does not address the probability of the harm. An exposed electrical wire will increase the chances of harm, but the hazard stays the same: an electrical shock could kill a person.

The importance of identifying hazards, therefore lies in the evaluation of their potential for causing or contributing to harm. While it is true that all workplaces have inherent safety hazards, the type and severity of these hazards vary with the work carried out in the environment. Settings like manufacturing plants, construction sites or forestry operations typically present more, and significantly greater, hazards compared to relatively safer environments like a corporate office.

The process of hazard identification in organisations is often similarly variable. There is a tendency in some organisations either to over-report minor hazards by reporting typical housekeeping issues as hazards, or to under-report major hazards due to a low perception of risk. Additionally, some organisations use hazard identification as a lead indicator of health and safety performance, assuming that reporting more hazards equates to a more proactive safety-management system. This is a misconception.

In reality, within any specific industry, there is a finite number of significant hazards, typically ranging between 30 and 40 types. These hazards can manifest differently across various parts of a business. For instance, the risk of falls can vary significantly: falling on the same level (commonly called 'slips, trips and falls'), falling from a two-metre ladder, and falling from a six-metre scaffolding. While all are fall hazards, they are very different.

Classifying hazards into categories such as physical, chemical, biological, ergonomic, and psychosocial is also a commonly used approach. However, the practicality of this classification can be debated. Each hazard, regardless of its category, is unique. Simply categorising them in classifications does not necessarily enrich the understanding of the specific risks or management strategies required.

The following is a typical list of hazards in an industrial environment. It is not intended to be exhaustive but rather to provide examples of a typical list of hazards.

- **Manual handling**: Manual handling resulting from load overweight, load oversize, poor grip, unstable load, posture, twist, duration and frequency of the task, cold muscles, lack of fitness, poor technique, etc. Musculoskeletal injuries.
- **Forklift operation**: Pedestrian within forklift operating area hit by a moving forklift or the load. At risk forklift operation (speed, turning with raised load, etc.) with injury to forklift operator or pedestrian. Environmental hazards (e.g., unguarded loading dock, uneven surface, conflicting traffic movement, etc.). Carbon monoxide (CO) exposure.
- **Noise (+ 85 dBA/8 hr or impulse noise + 140 dBA)**: Exposure to noise at work, including continuous noise (constant and stable over a period of time), variable or intermittent noise (fluctuates between quiet and loud over time), and impulse or impact noise (very high intensity and very short duration, e.g., explosion).
- **Objects falling from height**: Falling objects, e.g., tools falling from work platforms, unstable and over-stacking of goods, unsecured loads on vehicles, etc.
- **UV radiation**: Prolonged exposure to the sun or artificial sources of UV light. It can cause skin damage and cancer, eye damage and immune system suppression.
- **Elevated work (between 1 m/3 ft and 2 m/6 ft)**: Fall onto the floor level, e.g., ladder use, carrying items up/down a ladder, working on a ladder.

- **Elevated work (higher than 2 m/6 ft)**: Working at very high levels, e.g., on roofs, high scaffolding, exposed places, climbing up/down from an elevation, etc.
- **Fire**: Fire, either intentional, e.g., gas burners, or unintended fires, e.g., spontaneous combustion, accidental fires, etc.
- **Hot surfaces, objects, or steam**: Skin contact with boiling water, steam, or hot object/surface where temperature is greater than 60°C and duration of contact is more than 5 sec.
- **Travelling at speed, colliding with another object**: Include travelling in a car, boat, operating machinery, etc. Speeding reduces reaction time, leads to loss of control and colliding with a person(s), another vehicle, object or animal.
- **Working on or in close proximity to live electricity**: Person touching an exposed live part with a hand or tool. Can cause burns, shock trauma, ventricular fibrillation. Fire at circuit board.
- **Electrical fault (mains voltage)**: Fault in an electrical device, incorrect use of electrical device. Can result in burns, shock trauma, ventricular fibrillation, and a fire at circuit board.
- **Stress, bullying, violence**: Work-related stress, bullying, harassment, physical violence. Can result in psychological harm.
- **Biological hazards**: Contact with pathogens, e.g., contact with bodily fluids, airborne pathogens like influenza, tuberculosis, COVID-19, moulds, fungi, yeast, and potentially infectious waste.
- **Fatigue – a state of physical or mental exhaustion**: Diminished capacity for work and reduced efficiency, usually resulting from prolonged stress, overwork, or illness.

- **Cold stress**: Working in a cold climate. Frostbite to extremities, hypothermia.
- **Heat stress**: Working in a hot/humid environment or climate. Serious medical conditions can result.
- **Workstation ergonomics**: Poor posture, overuse, forceful gripping, rotation, working above shoulders or bending. Can cause carpal tunnel syndrome, tendinitis, rotator cuff injuries. muscle strains and low-back injuries.
- **Inhalation of dust, welding fumes & gases**: TIG/MIG/Arc welding of various metals. Sanding, sawing, grinding or various materials to create dust. Potential respiratory effects, depends on the dust toxicity and exposure.
- **Inadequate lighting and glaring**: Insufficient light, too bright, uneven light, back light, reflection. Can cause strain on eyes; unable to see existing hazards.
- **Hazardous chemicals**: Coming into contact/working near products or chemicals with properties that are explosive, flammable, oxidising, toxic, corrosive or toxic to the environment.
- **Confined or restricted space**: Enclosed or partially enclosed space not intended or designed primarily for human occupancy and may present harmful airborne contaminants, unsafe concentration of flammable substances, unsafe levels of oxygen, substances that can cause engulfment or simply lead to poor work posture due to limited space.
- **Vibration (hand, arm)**: Use of hand tools that vibrate, e.g., sanders, linishers, saws, impact drills, angle grinders, reciprocating saw. Hammering steel, e.g., to flatten or shape steel. Can result in fingers, hands tingling, numbness, loss of touch, reduction in grip. Hand pain. Nerve damage.

- **Sharp objects**: Handling sharps, e.g., sheet metal, equipment edges, tools, knives, etc.
- **Pneumatic/compressed air**: Compressed air that enters the bloodstream, is blown into the mouth or eyes. A ruptured hose can cause whiplash injuries.
- **Flying projectiles**: Use of equipment that produces dust/grit/swarf or cutting fluid splashes such as burnishing metal, grinders, manual lathe, power tools, saw bench, or blowing down with compressed air. Can cause penetrating injuries.
- **Rotating or moving parts**: A variety of mechanical motions and actions, including the rotating and reciprocating parts, moving belts, meshing gears, cutting teeth, and any parts that could impact or shear.
- **Working in water**: Hazards could include slips, trips and falls, particularly when wading or walking in the water, exposure to contaminated water, such as chemicals, substances, polluted, biological, etc., insect stings/bites, drowning, cold-water shock, and immersion.
- **Animals/insects**: Working/coming in contact with animals/insects, e.g., impact from animal, animal/insect bites, crush by an animal and exposure to zoonotic diseases.
- **Violence in the workplace**: Psychological and physical harm as the result of bullying and aggressive behaviour from co-workers, customers or the general public. Bullying could be blatantly obvious, or it could be very subtle and not obvious.
- **Natural hazards – weather**: This could include avalanche, flooding, cold/heat wave, earthquake, hail, etc.
- **Shift work**: Disturbance of the circadian rhythm could lead to sleep cycle disturbances and fatigue, an increased risk for accidents and work-related mistakes, impaired

cognition and decreased job performance, adverse health consequences, and mental health deterioration, including maintaining social and family involvement.

Some specialised situations, such as firefighting, underwater operations, working on fishing boats, oil rigs, etc., will undoubtedly pose hazards not listed. It is always advisable to involve multidisciplinary specialists to identify the special hazards in these cases.

Furthermore, each of these hazards may also be present at multiple places in a workplace but not always in the same way. For example, a motor vehicle travelling at 30 km/h poses a different hazard than the same vehicle travelling at 160 km/h. Working at heights close to overheard electrical lines will require different controls than other cases of working on scaffolding.

Understanding Hazard Severity Assessment

Identifying the hazards enables the organisation to assess the potential severity of these hazards. Most people do not find this assessment difficult; they instinctively understand that being hit by a slow-moving vehicle may only cause relatively minor injuries, whereas a vehicle travelling at 160 km/h will almost certainly kill the person.

The difference may not always be as big as the above example, but most of the time an assessment team can come to a consensus on what the level of potential harm is.

Levels of Potential Harm

The following five levels of potential harm caused by a hazard are commonly recognised and should be sufficient:

1. **Insignificant Harm (First Aid Injury):** Exposure to this hazard might result in minor injuries requiring only first aid treatment. Typically, the individual can immediately resume normal duties.
2. **Minor Harm (Medical Treatment Injury):** This level involves injuries that necessitate professional medical treatment. However, recovery is relatively swift, and the nature of the injury allows the person to return to work immediately or by the next rostered workday.
3. **Moderate Harm (Lost Time Injury):** Here, significant but non-permanent injuries or illnesses occur, necessitating absence from work for one or more shifts subsequent to the injury.
4. **Major Harm (Disability Injury):** Exposure could result in permanent impairment or life-changing injuries, like vision/hearing loss, amputation, spinal cord injuries, severe burns, or organ damage. Recovery, if possible, is often prolonged and complex.
5. **Catastrophic Harm (Fatality):** This is the most severe level, where incidents could result in the death of one or more individuals.

When the potential harm is known and the assessment team came to a consensus of the potential severity, the next aspect on the risk matrix can be assessed: the likelihood that the assessed level of harm will occur, the topic of the next chapter.

Controlling Hazards

It is a common misconception that administrative controls in the Hierarchy of Control can lower the severity rating on the Risk Matrix. This belief is incorrect. Administrative controls, by their nature, cannot directly mitigate the hazard itself; they modify only the risks associated with a particular hazard.

A hazard – any object, condition, or activity/inactivity – that could potentially cause injury or illness must be fundamentally altered to reduce its inherent severity. For instance, implementing a safe procedure when using a potentially harmful object does not diminish the object's capacity to cause harm. It merely lowers the likelihood of an injury occurring.

Despite its logical basis, this distinction is often overlooked, leading practitioners to inaccurately record a reduced severity rating after implementing only administrative controls. To genuinely lower the severity rating, the nature of the hazard must be changed, typically through engineering controls or other higher-order measures in the Hierarchy of Control.

This point is crucial and bears repeating: relying solely on administrative controls to lessen potential harm fundamentally weakens the integrity of the entire risk-control process. The hazard itself must be altered first to reduce the potential harm it can cause, and then all the possible controls in the Hierarchy of Control can be used to reduce the likelihood of the remaining potential to harm from occurring.

Consider a manufacturing environment. Engineering controls such as reducing water temperature from 100°C to 50°C alters the hazard and therefore reduces the potential severity. The same applies to electricity; changing the voltage from 240V to 24V would also reduce the potential severity.

In contrast, an administrative control would not alter the water temperature or the voltage. Instead, it might for example involve implementing a policy where workers are not allowed to enter an area where hot water is used or installing an RCD/RCCB circuit breaker to prevent a human from becoming exposed to the hazard. The hot water and the electricity could still seriously harm a person; however, the controls reduce the chances of that happening.

Note, this does not mean that all engineering controls will reduce the potential harm the hazard could cause. The distinction is that the hazard itself must be changed. If the engineering control does not change the hazard, but only adds something to it to reduce the chances of harm, then the potential harm is not affected. For example, adding a guard to a rotating machine does not alter the potential harm the spinning machine could cause; it reduces the opportunities for someone to become entangled in the machine.

To summarise, effectively mitigating the potential severity of a hazard often involves a combination of strategies, including the following. This is not an exhaustive list, but it illustrates the focus on reducing the potential severity of harm:

- **Redesign the Hazard**: Modify the object to remove or change hazardous features.
- **Hazard Substitution**: Replacing the hazardous element, whether it is an object, condition, or specific action/inactivity, with a safer alternative.
- **Process Modification**: Altering the sequence of operations or actions to eliminate or diminish the need for hazardous elements, thereby reducing the potential impact.
- **Restrictive Measures**: Enclosing hazardous objects, conditions or activities to prevent/limit access.
- **Safety Interlocks**: Implementing devices that prevent machinery from operating under dangerous conditions.
- **Technological Innovations**: Leveraging new technologies and innovations.
- **Ergonomic Adjustments**: Making ergonomic adjustments to workspaces and tools.

Chapter 8
Workplace Risks

Risk, in the context of workplace health and safety, is defined as the likelihood of harm or adverse health consequences resulting from exposure to a hazard. It is important to understand that risks are inherently tied to hazards – they represent the probability that these hazards could cause harm.

The Role of Risk in the Risk Matrix

In the Risk Matrix, the risk is categorised along the likelihood axis. This axis reflects the probability that exposure to a hazard will result in harm. Risk assessment is more nuanced than hazard assessment. As mentioned before, most people understand the levels of potential harm, such as when an injury requires first-aid treatment, versus having to visit a doctor. However, the chances that a person will place their hand inside a rotating machine is more subjective, making consistent and accurate evaluation of risk more challenging.

Currently, many organisations have one-line descriptions of likelihood, including phrases like "expected to occur frequently", "will probably occur in many circumstances", "might occur at some time", "could occur but is considered rare", and "not expected to occur". Consistently interpreting these terms is very difficult, and these short descriptions do not provide much guidance. It is even debatable if the spacing between the descriptions is equal; is the gap between "expected to occur" and "probably occur" the same as between "might occur" and "could occur"?

Critical Factors Influencing Risk Assessments

Safety 2.1 recognises this problem and the reality that there is no simple solution. However, it believes that using a multifaceted approach increases the quality of the decision. Instead of relying on a solitary, overarching statement, it introduces eight interconnected factors to assess likelihood. The interplay among these factors is crucial, and their collective impact should be considered to form a comprehensive assessment of likelihood. These factors are:

1. **Hazard Recognition**: This refers to the ease or difficulty with which a hazard can be identified in advance. Some hazards are readily apparent, such as the audible approach of a noisy vehicle, which provides clear, immediate sensory feedback. In contrast, other hazards are far less conspicuous. Electricity, for instance, gives no visible or audible warning to indicate whether a circuit is live or de-energised. This lack of sensory cues can significantly increase the risk associated with electrical hazards.
2. **Risk Isolation**: The ability to isolate a hazard effectively can greatly reduce the likelihood of harm. When hazards can be isolated, such as keeping a safe distance between

machinery and personnel, the risk is minimised. However, when isolation is impractical, such as needing to calibrate a machine while it is in full operation, the risk of harm is heightened due to the direct exposure to the hazard.

3. **Avoidance Capability**: This factor assesses the ability to avoid hazards. Some scenarios allow for quick reactive measures, such as moving away from an impending explosion, if there is an escape route and time to react. However, in situations where operators are confined or encaged, their capacity to avoid hazards is severely limited, elevating the risk.

4. **Ease of Control**: The degree of control over a hazard, as well as the ease with which these controls can be applied, is crucial in assessing the likelihood of harm. Automated controls significantly lower the likelihood of harm, such as a machine that only operates when all safety guards are engaged. In contrast, if controls require substantial human intervention or are difficult to activate, the risk of harm increases due to the potential for error.

5. **Frequency of Hazard Presence**: This involves the regularity with which a hazard is present in the work environment. Frequent exposure to a hazard typically increases the likelihood of injury, as the operator is more often exposed to the risk. Conversely, infrequent hazards can lead to complacency and reduced vigilance, potentially making rare but serious incidents more dangerous. This dual nature of hazard frequency must be considered in the context of when and where the hazard occurs.

6. **Duration of Exposure**: The length of time individuals are exposed to a hazard is another factor in risk assessment. Short-term exposure may reduce the likelihood of harm, particularly when combined with other mitigating factors, such as infrequent occurrence. For instance, a road worker

constantly near fast-moving traffic faces a higher risk compared to someone in a factory yard with only occasional vehicle movement. Duration of exposure therefore also plays a role in determining the likelihood of potential harm.

7. **Resource Availability**: This concerns the availability of necessary facilities, technology, tools and equipment to manage the hazard effectively. Inadequate resources can, for example, lead to improvisation, which often increases the likelihood of injury.

8. **Job Design Factors**: These are elements of job design, such as work sequencing, scheduling and dependence on preceding activities, that impact hazard exposure. Poor job design can significantly elevate risk. For example, working in confined spaces, under intense time pressure, or in poorly planned environments leading to operational bottlenecks can increase the likelihood of incidents.

Likelihood Ratings and Factor Definitions

Each of the above factors should be carefully evaluated against the 'Likelihood' ratings on the Risk Matrix, which range from 1 (Highly Unlikely) to 5 (Highly Likely). The following definitions for each level of likelihood may assist the final decision.

As mentioned before, the individual ratings on these eight factors are not the point, as they all provide information to make the overall likelihood rating on the Risk Matrix more accurate.

Having said that, the information should not be discarded either. Knowing what the main risk-contributing factors are will be valuable information when the risk is controlled. For example, if the problem is that the hazard is not obvious and very hard to identify in advance, risk controls may be aimed at making it more obvious, such as adding alarm systems.

Likelihood Rating (that harm at stated level will occur)	1 Highly unlikely	2 Unlikely	3 Possible	4 Likely	5 Highly likely
Ability to recognise the hazard	The hazard is obvious	Relatively easy to recognise the hazard in advance	Possible to recognise the hazard in advance	Difficult to recognise the hazard in advance	Almost impossible to recognise the hazard in advance
Ability to isolate the hazard	Easy to isolate the hazard	Most of the time relatively easy to isolate the hazard	Sometimes possible to isolate the hazard	Difficult to isolate the hazard	Almost impossible to isolate the hazard
Ability to avoid the hazards	Easy to avoid the hazard when it occurs	Most of the time relatively easy to avoid the hazard	Sometimes possible to avoid the hazard, but not always	Difficult to avoid the hazard when it occurs	Almost impossible to avoid the hazard when it occurs
Ease of Control	Controls are obvious and easy to implement	Controls are most of the time at hand, relatively easy to implement	Controls are not always at hand, and/or somewhat difficult to implement	Controls are either complicated and/or difficult to implement	Very difficult to control the hazards; requiring extensive expertise
Frequency of hazard's presence	The hazard is very rarely present	The hazard is only at limited times present	The hazard is sometimes present	The hazard is regularly present	Hazard is ever-present
Duration of exposure to the hazard	Almost never exposed to the hazard	Exposed to the hazard for short periods of time	Exposed to the hazard for medium periods of time	Exposed to the hazard for relatively extended periods of time	Almost always exposed to the hazard
Availability of facilities, tools/equipment, etc.	Facilities, tools technology, and equipment are available	Most facilities, technology, tools, and equipment are available	The key facilities, technology, tools, and equipment are available	Only limited, tools facilities, technology & equipment available	No facilities, technology, tools, and equipment are available
Job design, e.g. sequencing, resourcing, scheduling, etc.	Work is optimally designed, no limitations on production process, resources allocations, scheduling, etc.	In most cases the work is planned to optimise work sequencing, allocation of resources, and scheduling	It is sometimes not possible to avoid unplanned work, limited resource availability, tight scheduling, etc.	Difficult to avoid unplanned work, limited available resources, or very tight scheduling.	Almost impossible to avoid unplanned work, working only with limited resources, urgent jobs, etc.

Figure 7. Range of Likelihood Ratings.

Risk Controls

Controlling risk is not straightforward. Not all controls work equally well, and safety practitioners sometimes under- or over-estimate the effect controls may have on the risk. Furthermore, although risk controls span the entire Hierarchy of Control, they often lean heavily on what is known as administrative controls. These include safety procedures, meetings, scheduling, shift rotations, personal protective equipment (PPE), emergency plans, safety signs, access restrictions, and maintenance. However, with a few exceptions, these mainly aim to modify human behaviour rather than tackle the hazard directly. There is typically a very heavy reliance on safety procedures like safe operating procedures (SOPs), safe work method statements (SWMS), and job safety analyses (JSAs). Perhaps a more fitting name for administrative controls would be 'human controls'.

Safety 2.1 criticises over-reliance on procedures. Traditionally, these procedures dictate the safest way to perform tasks, a concept rooted in Safety 1. This approach often leads to stricter regulations following incidents, aiming to correct or prevent deviations from the procedures. Yet, this approach may lead to more prescriptive procedures that, ironically, are increasingly ignored by frontline operators in a complex adaptive system.

In contrast, Safety 2.1 follows a different approach: employ all available controls to bring risk down to an acceptable level, resorting to procedures only as a last resort. The idea is to prescribe just enough to manage the risk, leaving room for frontline workers to apply their knowledge and skills within what is called the 'safety envelope'. This concept encourages a balance between control and flexibility, enabling workers to determine the safest course of action within a defined risk threshold. Part Three of the book will explore the safety envelope in more detail.

Word of Caution

A serious word of caution: Do not use these eight factors to develop a mathematical calculation of pre- or post-control risk. Even worse, do not develop a software solution to calculate the risk. Safety 2.1 explicitly opposes linearity; hazards are complex systems, and the influence of any one factor on the level of risk varies each time a risk presents itself. Integrating these factors into an overall risk rating and subsequently developing controls are expert tasks that should not be delegated to non-human artificial intelligence.

Additionally, these factors are not an exhaustive list of potential elements in understanding risk. For example, a switching schedule when energising an electrical network does not fit neatly into any of these factors, even though many of these factors will either explicitly or implicitly be included. However, a well-developed switching schedule, supervised step-by-step by an external party, such as a control room, is an irreplaceable control when operating switchgear in a substation.

Chapter 9
Spider Diagram

This chapter introduces a visual tool designed to simplify the risk-control process. It predominantly focuses on the likelihood aspect of risk control. As elaborated in Chapter 7, the only way to minimise the potential severity of harm that a hazard could cause is by altering the hazard itself, which is fundamentally an engineering challenge. However, reducing the likelihood of harm can not only be achieved through engineering solutions but often requires administrative measures as well. The Spider Diagram has been developed specifically to facilitate the management of these more variable controls.

It is important to clarify that emphasising the likelihood factor through this tool does not diminish the significance of hazard severity. Indeed, when measures aimed at lowering or eradicating the potential for serious harm prove effective, the consideration of 'likelihood' naturally becomes less critical.

The tool proves especially valuable when modifying the hazard did not sufficiently lower the risk score on the Risk Matrix below an acceptable threshold, as discussed in Chapter 7.

Spider Matrix Template

The template below is used to record the risk assessment and the impact that various controls could have on the likelihood. It is not so much intended to be a reporting tool but rather a visual guide for the team doing the risk assessment.

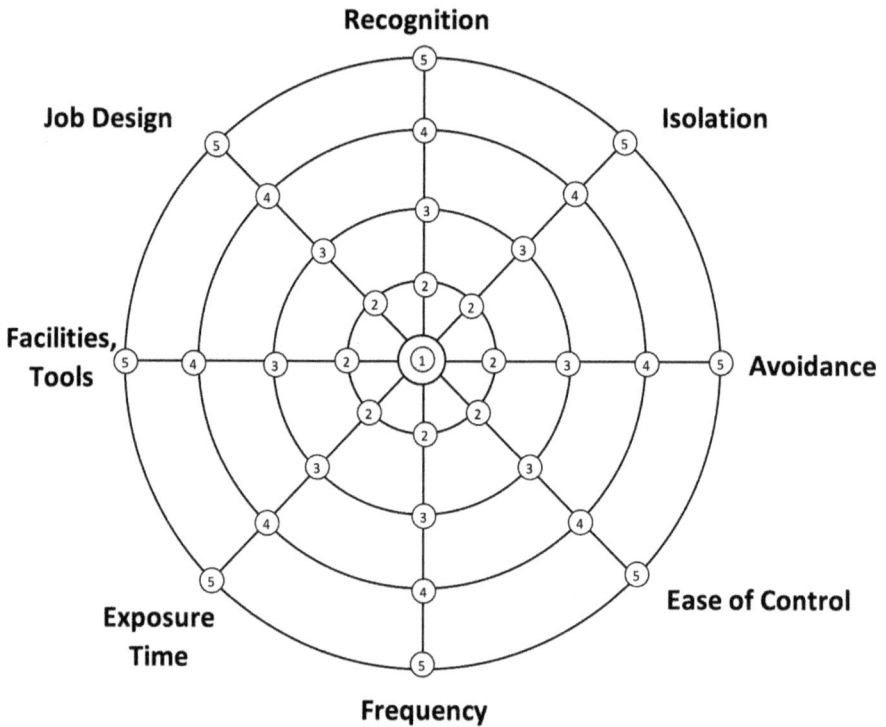

Figure 8. Spider Diagram.

The Spider Diagram visually records the individual ratings of the eight factors contributing to the likelihood of harm, as discussed in the previous chapter. This offers the opportunity for the assessment team to consider each of the assessments but, importantly, allow them to consider the interplay between these factors. The factors do not only individually contribute to the likelihood of harm; they also often combine with other factors, increasing or decreasing their impact on the overall level of likelihood.

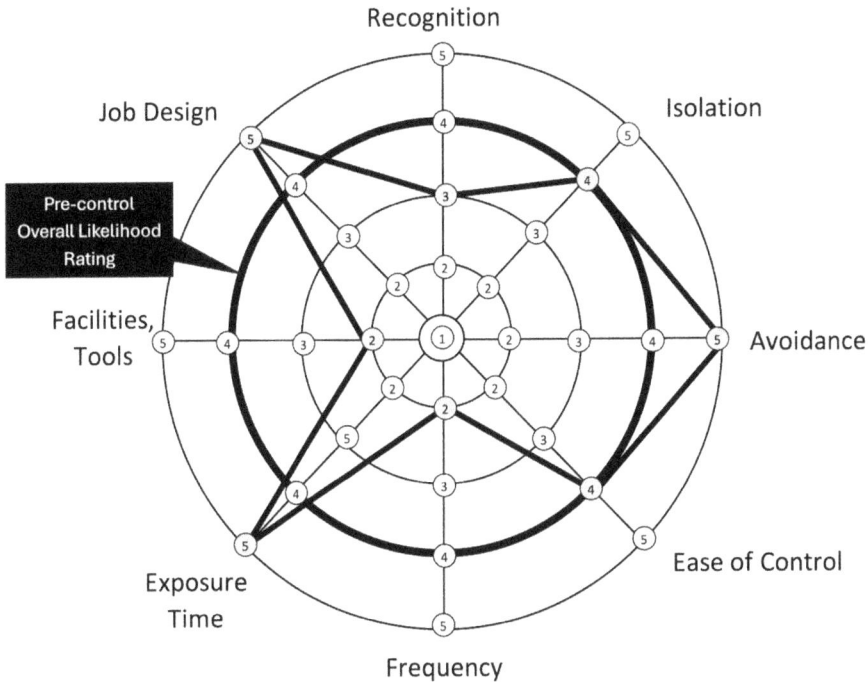

Figure 9. Spider Diagram: Eight Factors and Overall Rating.

Likelihood Reduction

The topic of likelihood reduction, also covered in Chapter 8, presents a challenge. One common issue is the overestimation of the effectiveness of controls in reducing the likelihood of a hazard. In other instances, less obvious factors may be overlooked, leading to a reliance on procedures and instructions as their primary control mechanisms. This is indicative of a traditional safety management response, which often fails to consider the broader spectrum of control options.

Safety 2.1 advocates for a more expansive approach to identifying potential controls. This involves considering controls to each of the eight factors contributing to the overall likelihood rating, as well as the interplay between these factors. For example, if it is difficult to avoid a hazard, but it is possible to make recognising the hazard earlier, it could create more opportunities to avoid the hazard.

By addressing these aspects, Safety 2.1 moves beyond traditional safety practices, promoting a more holistic and effective approach to reducing the likelihood of hazards in the workplace. This shift involves a thorough assessment of work processes, hazard characteristics and control mechanisms, ensuring a comprehensive strategy for risk mitigation.

Navigating Control Effectiveness

The Spider Diagram can also be used to indicate the amount of control that is required by adding the maximum likelihood rating the organisation will tolerate, as discussed in Chapter 6. In the example below, the likelihood rating must be reduced from 4 (likely would happen sometime) to 2 (rare, conceivable but unlikely).

Assessing the effectiveness of controls to achieve this required reduction can be tricky. Sometimes the controls' impact is overestimated, while in rare cases, they are underestimated. Unfortunately, there are no hard-and-fast rules for this; it ultimately comes down to professional judgement.

Here are some 'stronger controls' that typically lead to a reduction in risk likelihood on the Risk Matrix by one rating scale point. For instance, they might shift the assessment from "Highly likely/expected to happen" to "Likely would happen sometime," or from "Might happen at some time" to "Conceivable, but unlikely".

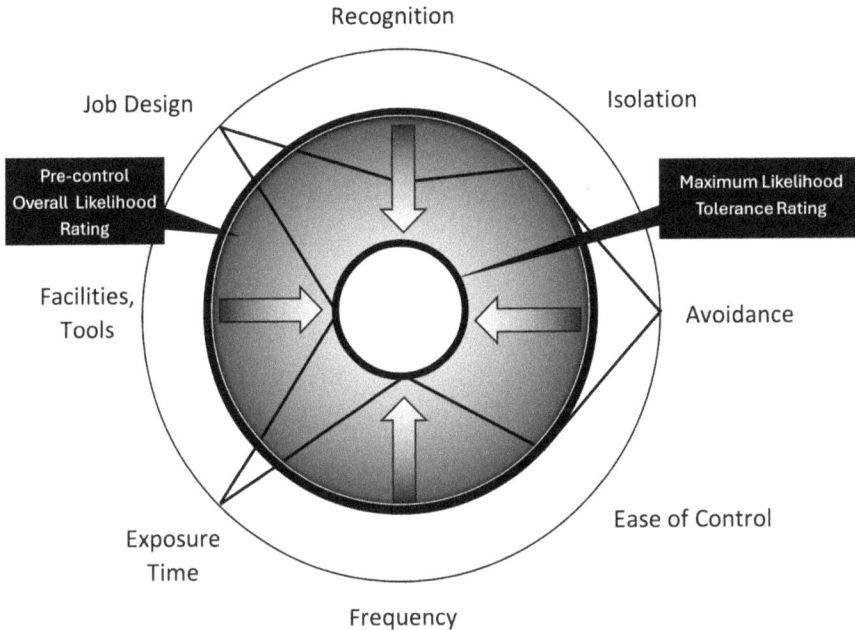

Figure 10. Spider Diagram: Risk Mitigation.

As a rule of thumb, administrative controls will very seldom move more than one scale point on the Risk Matrix:

- Making hazards more visible by using spotlights or other means
- Installing guards to prevent operators from getting entangled
- Increasing the distance between the hazard and the operator
- Adding handles or stabilising devices to prevent sudden movements
- Adding warning signals when a hazardous condition is imminent
- Reducing the frequency or duration of hazardous activities
- Implementing monitoring (e.g., noise, chemical, radiation) to assess exposure levels

- Providing specialised tools for unique tasks
- Using personal protective equipment (PPE) designed for specific risks
- Having a control room operator confirm each step during high-risk operations, i.e., second person oversight
- Introducing preventative maintenance to increase machine reliability, and regularly rotating tasks to mitigate fatigue.

On the other hand, 'weaker controls' alone do not significantly reduce risk on the Risk Matrix. They require combination with other controls to warrant movement in risk assessment. Examples include:

- Safety signs (they do not independently shift risk levels)
- Poorly designed safety instructions (only providing general guidance)
- Overly complicated procedures (making their use onerous and impractical)
- Safety talks and alerts (not making a direct impact on the risk)
- Wearing standard safety gear (like hi-vis vests, safety shoes, and hardhats) without specific risk-mitigation benefits.

The space inside the maximum tolerance level on the Spider Diagram, the centre of the Spider Diagram, is an interesting space and the topic of the next part of the book.

Part 3
The Safety Envelope

Chapter 10
Safety Envelope Defined

It is sensible to consider the bigger picture before defining what a 'safety envelope' is.

In many legal frameworks globally, safety legislation dictates that organisations should reduce risks to a threshold defined as 'As Low as Reasonably Practicable' (ALARP). This concept, though deceptively simple in its phrasing, is frequently misinterpreted in its practical application. The legal definition of 'reasonably practicable' seems to conflict initially with the notion of an acceptable level of risk. It seems to imply that efforts to mitigate risk should continue until every possible control has been applied.

However, this interpretation misses the complexities inherent in the concept. Regulatory authorities across most jurisdictions understand that it is not only about eliminating risk, it is in many cases also about reducing the risk to a level that is considered tolerable. This concept is visually represented in the diagram from a WorkSafe New Zealand publication.

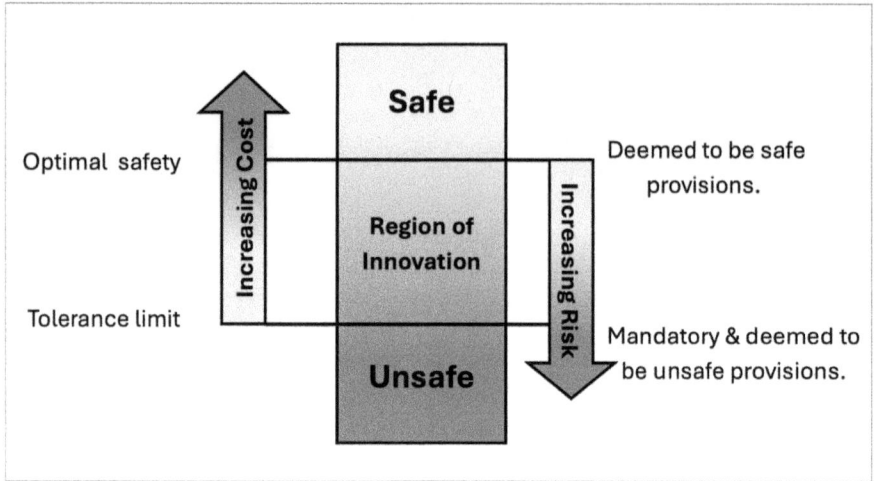

Figure 11. WorkSafe NZ Risk Tolerance.

Figure 11 illustrates that achieving a level of risk tolerance does not mean the absolute absence of risk. There often are residual risks that may not be addressed by standard control measures. This understanding is crucial to the development of the Safety 2.1 methodology, which introduces the notion of the 'safety envelope'.

The safety envelope is that space on the risk continuum where formal risk controls are not applied. It is a dynamic space where operators have the autonomy to make informed decisions and handle hazardous conditions effectively. Within this envelope, risk levels have already been controlled to fall within the predefined 'risk tolerance' range, primarily through engineering and, in certain instances, critical controls and processes.

This concept is fundamental to the Safety 2.1 framework. It denotes a significant departure from conventional safety methodologies, which typically rely on rigid, prescriptive guidelines for employees. The Safety 2.1 approach pivots towards empowering frontline workers, enabling them to evaluate situations and take necessary actions to manage risks inside a specific demarcated area – the safety envelope.

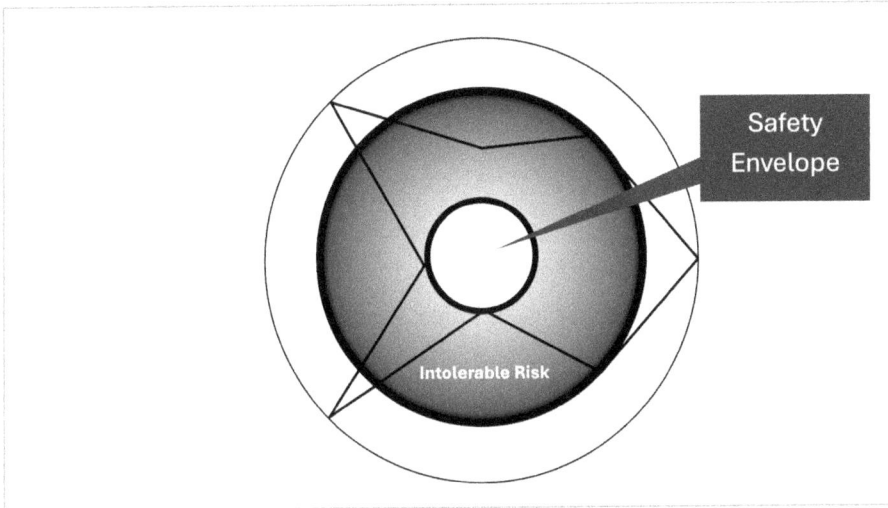

Figure 12. The Safety Envelope.

This 'safety envelope' transforms the aspirational goal of empowering workers in the Safety 2 ideology into a tangible and practical model. The approach moves away from a scenario where an organisation attempts to micro-manage every aspect of work through an exhaustive array of rules, procedures and processes. Instead, it delineates a clear demarcation point where the responsibility for safety decision-making transfers to the frontline operators.

However, this shift in approach does not suggest that organisations relinquish their duty of care towards their employees. Rather, it underscores a commitment from the organisation to provide its workforce with the necessary resources, skills and environment to make informed and timely decisions in the field. This includes creating a physically and mentally safe working environment, imparting essential skills and knowledge, cultivating a supportive and proactive organisational culture, and ensuring overall mental well-being in the workplace.

Adopting this new paradigm might necessitate a substantial shift in organisational thinking. It embodies the belief that 'we can trust our workers to execute their tasks competently and safely'. For many advocates of the Safety 2 philosophy, this trust in workers' competence and judgement is a core principle. It acknowledges that while human errors are inevitable, the ownership of actions enhances the chances of learning from mistakes. It also allows for leveraging the benefits of a complex adaptive system.

Safety 2.1 advances this concept by operationalising the theoretical underpinnings of Safety 2. It not only acknowledges the inevitability of human error but also recognises the value of learning from these errors and adapting processes accordingly. This operationalisation is a key step in translating Safety 2's theoretical framework into a practical, applicable model that can be integrated into everyday safety practices within organisations. This integration is aimed at creating more dynamic, responsive and resilient safety management that values both risk reduction and the empowerment of frontline workers.

Chapter 11

Operating inside the Safety Envelope

What happens inside the safety envelope is at the core of the Safety 2.1 approach. To recap, the safety envelope is the safety actions over which the worker will have independent control. The size of the envelope under the worker's control is limited by the risk controls identified and implemented as part of the risk assessment. These mainly engineering controls and non-negotiable administrative controls fall outside the 'safety envelope' and the worker has no choice other than to implement and maintain them. For example, if a guard has been installed on a rotating lathe and a non-negotiable critical procedure has been introduced, the worker cannot remove the guard or ignore the procedure.

Although these 'compulsory' controls have already reduced the risk to within the risk tolerance levels, it does not mean all risk is removed. On the contrary, what happens in the remaining safety envelope will not only ensure the risk is even further reduced, but it will also be a significant factor in ensuring the engineering controls and non-negotiable administrative controls are adhered to and maintained.

The size of the safety envelope varies based on the risk level associated with a specific hazard that the organisation has accepted. For instance, if the organisation deems a risk intolerable and formally mitigates it, the resulting envelope is smaller. In this case, frontline operators have limited discretion in dealing with remaining risks. Conversely, if the organisation chooses not to formally mitigate risks and tolerates them, the envelope expands. Frontline operators then have more autonomy in deciding how to handle these risks.

Although what happens inside the safety envelope is left up to the worker to determine, it does not imply that workers are left to their own devices. The organisation has four specific areas where it can contribute to the worker's success in the envelope; these are providing a safe working environment and the appropriate tools, ensuring workers have the required skills and knowledge to perform the work safely, influencing the safety cultural behaviour patterns of the workers, and ensuring workers have the mental ability to perform the work safely.

Working Environment and Tools

The work environment plays a crucial role in ensuring safety as it directly influences the behaviour, health and well-being of employees. A well-maintained, ergonomically designed and hazard-free workplace not only minimises the risk of accidents and injuries but also boosts morale and productivity, as employees feel valued and secure.

A positive environment with clear safety protocols encourages adherence to safety practices and fosters a culture of caution and responsibility. Moreover, a safe work environment is essential for compliance with legal and regulatory standards, helping to avoid legal liabilities and safeguarding the organisation's reputation. In essence, a safe work environment is foundational to the overall success and sustainability of any organisation.

The same applies to tools. Providing the frontline operators with all the tools and equipment they need to perform the work, and ensuring these resources are well-maintained, is crucial to performing the work safely.

Skills and Knowledge

For people to operate effectively inside the envelope, they must have the knowledge and skills required to perform the work. A lack of either knowledge or skills will impede their ability to take the best actions to mitigate risk.

First, this requires determining what knowledge and skills are required to perform the work, then an assessment of the operator's current abilities against the requirements. The gaps then need to be closed through learning and development opportunities.

Regulatory requirements, official information resources, codes of practice, as well as licensing and qualifications requirements, must be included when upskilling the workers to do the work safely.

Safety Culture

Safety culture refers to the shared values, beliefs and norms of a group. This influences the ideas, customs and social behaviours as well as the way group members act.

It implies that a group of people will typically do things similarly and think similarly about a given topic. This has a profound impact on how workers will operate inside the safety envelope. For example, if the group 'approves' of taking short cuts when performing a certain task, the group members will most likely take these short cuts if not closely supervised. Conversely, if the group's cultural behaviour emphasises performing work safely, no outsider will have to tell them to do so.

Mental Ability

The mental ability of workers is influenced by the less obvious and more subtle organisational climate. This reflection of the collective mood and perceptions in a workplace is shaped by leadership's communication style, motivational strategies and daily interactions. It is also influenced by access to company information, the lack of work obstacles, organisational structure and how employee contributions are recognised. This dynamic climate essentially shapes an employee's experience and perception of the work environment.

All four of these areas are important and will be discussed in the following chapters. Safety culture and mental health in particular are often misunderstood and will receive specific attention.

Chapter 12
Work Environment and Tools

The work environment is pivotal to ensuring that workers can safely operate inside the safety envelope. An effective workplace design that incorporates such things as ergonomic layout, access to quality tools and equipment and the like will enable the workers to operate safely inside the safety envelope as well as fostering a sense of pride and responsibility that encourages them to be vigilant and proactive in identifying and addressing potential hazards.

Work Environment

The following will typically contribute to a work environment conducive to safe operations.

1. **Workplace Layout and Design**: The physical layout and design of a workplace have a significant impact on safety. This includes ensuring a logical flow of production, establishing clear exclusion zones for hazardous activities, installing barriers to separate different work activities, and providing ample space for safe movement. The layout

should be reviewed regularly to identify potential safety hazards and to adapt to changes in work processes or equipment.

2. **Electrical Safety**: Ensuring the safety of electrical systems through regular inspections and maintenance is vital to prevent hazards such as electrical shocks or fires. It is crucial to have marked emergency shut-off points and to ensure all equipment is properly grounded.

3. **Traffic Control and Safety**: Managing traffic flow in areas where vehicles or heavy machinery are in use, such as warehouses or construction sites, is vital. Establishing clear rules and pathways for vehicle movement helps prevent accidents. Key measures include creating designated pedestrian zones, mandating the use of high-visibility clothing for workers exposed to the vehicles, and implementing effective signalling systems to guide vehicle movement.

4. **Safe Handling and Storage of Hazardous Materials**: Standardised methods for the handling, storage and disposal of hazardous materials are crucial. This includes sufficient storage capacity facilities, meeting the safety data sheet (SDS) prescriptions, and clear labelling, to mention a few.

5. **Temperature and Humidity Control**: In workplaces like kitchens or industrial facilities, or working outdoors, where temperature and humidity levels can pose safety risks, maintaining a comfortable and safe physical environment is key. This includes measures to prevent heat-related illnesses and ensuring that high humidity levels do not create hazardous conditions, like slippery floors. Regular monitoring and adjusting of environmental conditions, as well as providing personal protective equipment (PPE) suitable for these environments, are important.

6. **Natural Light and Views**: Access to natural light and views of the outdoors can greatly enhance employee well-being, reduce stress levels, and contribute to a safer work environment. Designing workspaces to maximise natural light exposure and providing areas where employees can enjoy outdoor views during breaks can improve overall workplace safety and morale.

7. **Rest Areas and Break Rooms**: Having dedicated spaces for employees to take breaks is important for reducing fatigue and maintaining health. Comfortable and well-equipped rest areas and break rooms can provide a much-needed respite from work, helping to rejuvenate employees and maintain high levels of safety and productivity.

8. **Sanitation Facilities**: Providing adequate sanitation facilities, including restrooms and access to clean drinking water, is fundamental to maintaining hygiene and health in the workplace. Regular cleaning and maintenance of these facilities, along with easy access for all employees, are essential for a healthy work environment.

9. **Health and Safety Signage**: Posting clear, visible signs that identify hazards, instruct on safety protocols, or guide during emergencies enhance safety awareness. Regular updates and audits of signage to ensure relevance and visibility can further improve safety communication.

10. **Effective Housekeeping**: Maintaining a tidy workspace, devoid of clutter and debris, can contribute to minimising physical hazards. Strategic placement of adequate waste-disposal stations throughout the work area significantly contributes to this goal. By focusing on these areas, organisations can create a physical work environment that not only minimises the risk of accidents and injuries but also promotes a culture of safety and health awareness.

Tools and Equipment

The same applies to providing the frontline operators with the means to perform the work. This is in today's electronic age even more important as technology changes very rapidly and it puts an even larger responsibility on the organisation to provide the required tools and equipment. The following are examples of this:

1. **Tools of the trade**: Providing the operators with good-quality tools to perform their work with. This could include hand tools, specialised devices, and other objects appropriate for their tasks.
2. **Equipment**: This includes all equipment and systems that facilitate effective, and therefore safe, work. It could be large and small: lathes, work benches, presses, drills, and a myriad other equipment, as applicable to the work they perform.
3. **Technology**: Providing the operators with advanced technology, rather than expecting them to use old-fashioned and outdated techniques, will have a significant impact on safety performance.
4. **Data and Information**: The same applies here. If the operator has good data and information available, they have to rely less on inaccurate data or guesswork, reducing the risk of mistakes that could negatively impact on their own or others' safety.

Chapter 13
Vocational Training

Vocational training is specialised education designed to equip individuals with the specific skills, knowledge and abilities necessary for performing tasks. This training, typically less theoretical and more hands-on, focuses on preparing trainees for specific trades, crafts or career paths. These skills are, in turn, cornerstones of performing work safely.

The following are key components of effective vocational training:

Sound learning principles: Adult learning is a specialised science and very different to how children learn. Andragogic principles emphasise self-directed learning, leveraging learners' experiences, relevance to professional contexts, readiness to learn, a problem-centred approach, and internal motivation for adult learners.

Curriculum Design: The curriculum should typically be designed in collaboration with industry experts to ensure that training remains relevant to the current needs of the industry. It must regularly be updated to keep pace with technological advancements and changing industry standards.

Skill Development: The primary focus is on developing practical skills essential for specific jobs. For instance, in manufacturing, these could include machine-operating skills, welding, or assembly-line work whereas in the automotive industry, it might involve engine repair, electrical systems, or bodywork.

Workplace Readiness: Alongside technical skills, vocational training incorporates workplace readiness. This includes safety protocols, understanding industry-specific regulations, and developing soft skills like teamwork, problem-solving and communication, which are crucial to a collaborative industrial environment.

Technological Integration: Vocational training increasingly incorporates modern technologies. For instance, trainees might learn to operate advanced machinery or use software for design and planning.

Hands-On Learning: Vocational training is heavily centred on practical, hands-on experiences. Trainees might work on real-world projects, use industry-standard equipment, or engage in simulations that mimic actual job conditions. This approach helps learners to gain a realistic understanding of the work environment and the challenges they might face.

Certification and Accreditation: Many training programmes offer certifications or accreditations that are recognised within the industry. These credentials can be crucial for employment, as they demonstrate a certain level of achievement.

Apprenticeships and Internships: Often, vocational training includes components of apprenticeships or internships, where learners get the opportunity to work under the supervision of experienced technicians. This not only provides real-world experience but also helps in building a professional network.

Adaptability and Continuous Learning: Industrial settings are dynamic and feature continuous advances and change. Vocational training therefore often emphasises the importance of adaptability and the need for continuous learning and skill upgrading.

Planned Refresher and Skills Assessment: Regular and planned refresher training is an integral part of skill development. As well as ensuring that employees are aware of changes in industry-specific regulations and other changes, it also provides the opportunity to sharpen the employees' less-used skills and increase their awareness of safety and technical requirements.

In summary, structured and well-planned vocational training will provide workers operating inside the safety envelope with the knowledge and skills required to make the correct work method decisions. This is often not a one-stage process, where all employees have the same latitude to make choices inside the safety envelope. Individuals may have different levels of proficiency and experience, resulting in varying levels of worker abilities. For example, disparate individuals might be classified as: 'unqualified, need supervision'; 'qualified, work under peer supervision'; 'qualified, work unsupervised'; and 'qualified, can provide peer supervision'. Individuals will then be allowed more independence, based on their knowledge and skills profile.

Chapter 14
Cultural Behaviour

Shaping the cultural behaviours of workers is a fundamental aspect of Safety 2.1. However, this topic is frequently misunderstood. More crucially, it is not unusual for safety professionals to grapple with understanding the role of culture in preventing injuries and the mechanics of how culture functions. While the term 'culture' is often tossed around in discussions, it is seldom accompanied by a substantial and meaningful definition.

The concept of 'behaviour' in the context of safety can generally be bifurcated into:

Task-centric behaviour resulting from an individual's knowledge and expertise. For instance, without possessing the requisite electrical acumen and skills developed through trade education and supplementary competence training, an individual would be incapable of safely executing electrical tasks.

Behaviour indicative of an individual's intrinsic approach to tasks. Such behaviour originates from personal and communal values, norms and attitudes, commonly referred to as culture.

While important, the focus in this chapter is not on the acquisition of knowledge and skills, as these are governed by adult learning principles and discussed in the previous chapter – Chapter 13. Instead, the focal point is the behavioural manifestation of culture.

Again, the term 'culture' is often used without defining what it means. Here culture is defined as follows:

> Culture is a comprehensive construct that refers to the collective manifestation of human creativity, influenced by the shared values, beliefs and norms of a group. It encompasses both tangible artefacts, such as art, music, architecture and manufactured goods, and intangible elements like ideas, customs, social behaviours and patterns of interaction.

It is important to note that this definition is focusing on a group of people, not on the actions of individuals, even though groups comprise a collection of individuals.

The concept of 'safety culture' can be inadvertently restrictive, suggesting it is distinct from broader cultural dynamics. In reality, the culture of various groups encompasses safety-related elements, both tangible and intangible. To fully grasp the construct of culture, it is imperative to consider it in its entirety, recognising its comprehensive impact beyond safety parameters.

The components of culture reflect and shape the identity and character of a group. Importantly, individuals often belong to multiple groups with distinct cultural identities, which can lead to cultural dissonance as differing cultural influences interact within a person's experience.

Culture, according to McDougall et al.,[*] is "a product of the network of symbols and meanings that cultural members negotiate, produce and reproduce over time". The way it changes is complex; attempts to intervene in a group's cultural behaviours may at times lead to very unpredictable responses.

Behavioural Functionality

The workings of culture are often misconstrued or overly simplified. The following sections will delve into the core components of behaviour and its development.

However, this book does not offer an exhaustive theoretical examination of culture; rather, it serves as a starting point for students deeply interested in this intricate subject. The focus here is on how culture bridges the gap between theoretical concepts and practical application, and serious scholars may want to explore the underlying theories further.

Nonetheless, to grasp the significant role that safety culture plays inside the safety envelope, and to understand how it can be shaped and influenced, it is crucial to explore some of its key aspects and operational dynamics.

Understanding Automaticity

People often act without conscious deliberation, their behaviours triggered by a myriad of circumstances. When a specific behaviour is regularly repeated, it evolves into an automatic response, akin to an instinct. These repetitive actions, more intricate than mere

---

[*] McDougall, M., Ronkainen, N., Richardson, D., Littlewood, M. & Nesti, M. (2020) Three team and organisational culture myths and their consequences for sport psychology research and practice, *International Review of Sport and Exercise Psychology* 13(1): 147–162.

habits, often encompass a series of actions, including decision-making processes, which is here called 'behaviour patterns'. Over time, these behaviour patterns embed themselves subliminally in our daily routines, requiring minimal conscious effort to put themselves into action.

John Bargh* characterises this phenomenon as the four horsemen of automaticity: lack of awareness, lack of intentionality, mental efficiency, and difficulty to control or stop a process. This automaticity often operates without our awareness, leading to a blurred line between unconscious behaviour patterns and conscious mindfulness. The brain's ability to transition swiftly between these states can be misleading by creating the illusion that the behaviour was conscious, while in reality no active thinking was involved.

This concept is a fundamental aspect of Safety 2.1 and challenges commonly held assumptions. Contrary to popular belief, human decisions and actions are not always the result of deliberate thought. While our capacity for thought distinguishes us from other species, our behaviours frequently follow existing patterns, influenced by Bargh's 'four horsemen'. In repetitive tasks, we often exhibit a lack of awareness and intentionality, decreased mental efficiency, and difficulty in controlling or stopping the behaviour.

Safety 2.1 diverges from the traditional view that behavioural change requires conscious thought and commitment. As will be explored in subsequent sections, much human behaviour operates in the 'automatic mind' and is influenced by automaticity. While quantifying this is difficult and not very helpful, automatic

---

* Bargh, J. A. (1994). The four horsemen of automaticity: Awareness, intention, efficiency, and control in social cognition. In Wyer & Srull (Eds.), *Handbook of social cognition: Basic processes; Applications* (pp. 1–40). Lawrence Erlbaum Associates, Inc.

behaviour could anecdotally be estimated to make up upwards of 80% of all behaviour in a typical repetitive-action job.

Recognising this reshapes our approach to behaviour change in the realm of safety.

Role of Cognitive Processing

This does not mean cognitive thinking plays no role. Although patterns devoid of conscious thinking dominate work behaviour, there are scenarios where conscious decision-making is vital, especially during unforeseen events. Cognitive processing demands more mental resources, and the brain tends to favour unconscious patterns over mindfulness; it is less energy-sapping. Nevertheless, the brain can swiftly transition between automatic and conscious modes; when an appropriate automatic behaviour pattern is not readily available or appropriate and conscious thinking is required to resolve the issue at hand.

Collective Behavioural Patterns

Anita Woolley et al.[*] have shown that groups exhibit collective intelligence, suggesting that group behaviours also follow patterns. Or, as Stalinski[†] puts it, "the potential for organisational cultures to understand and utilise both conscious and unconscious aspects of their collective mind and intelligence for positive benefit".

Group dynamics reinforce these patterns, often covertly. Respect for influential group members leads to emulation of their

[*] Woolley, A. W., Chabris, C. F., Pentland, A., Hashmi, N., & Malone, T. W. (2010). Evidence for a collective intelligence factor in the performance of human groups. *Science* 330(6004): 686–688.
[†] Stalinski, S. (2004). Organisational intelligence: A systems perspective. *Organisation Development Journal* [serial online]. 22: 55–67.

behaviours. There is also at times a level of coercion, where these 'elders and betters' in the group may enforce specific behaviour patterns.

Closely aligned is the fact that humans are 'herd animals', exhibiting "a form of convergent social behaviour that can be broadly defined as the alignment of the thoughts or behaviours of individuals in a group (herd) through local interaction and without centralised coordination".*

This tendency to follow the herd stems from a deep-seated need for belonging, often compelling individuals to conform to the group's norms and practices. As a result, group behavioural patterns frequently override personal patterns, as the desire to be part of the group dominates individual tendencies. Those who deviate from these group patterns – often labelled as mavericks, nonconformists, or individualists – tend to find themselves at the margins of the group, as their behaviours and attitudes do not align with the collective norms. This dynamic emphasises the strong influence of social conformity and the challenges faced by those who resist such conformity.

The reality is that this collectivism is often not resisted; being inherently social, humans tend to align their behaviours with their groups, prioritising group norms over personal inclinations.

Operational Versus Leadership Roles

The influence of behavioural automaticity differs markedly between various roles. In operational positions, the nature of tasks is usually routine, encouraging the formation of automatic behavioural patterns as a result of repetition. In contrast, leadership roles

---

* Raafat, R. M., Chater, N., Frith, C. (2009). Herding in humans. *Trends in Cognitive Sciences* 13(10): 420–428.

demand frequent, innovative decision-making, where automatic responses may not be adequate or effective.

Understanding this contrast is vital in safety management. Individuals in high-risk environments often operate within established behavioural patterns, utilising less active cognitive processing. Therefore, the development of safety processes and procedures must acknowledge and address this tendency towards habitual behaviour over cognitive deliberation. Such an approach will ensure that safety measures are effectively aligned with the behavioural tendencies of the target population, enhancing both the observance and the effectiveness of the measures.

Decision-makers, purely as a result of the type of jobs they perform, are less exposed to pattern-forming behaviour and may overestimate the role of cognitive thinking in the day-to-day activities of frontline operators. They may therefore often focus too much on cognitive thinking when attempting to change behaviour patterns.

Pattern Attributes

Behavioural patterns can encompass both beneficial and detrimental elements. An unsafe action, when repeated frequently, becomes ingrained. Subsequently, such actions are carried out automatically, without conscious awareness.

It is counterproductive to label patterns as 'good' or 'bad' because if it is embedded in the behavioural pattern, it is bound to recur. Additionally, the human brain's capacity for deductive reasoning – where specific conclusions are drawn from general principles – allows for pattern generalisation, which sometimes results in inappropriate pattern application in new situations. Conversely, if a pattern proves effective in a new context, it might be adopted for other marginally similar scenarios.

This inclination to adhere to ingrained behavioural patterns can therefore occasionally lead to mistakes, especially when the prevalent pattern is not suitable for a specific situation. To an outsider, such errors might appear as glaringly obvious and potentially indicative of a deliberate disregard for safe practices. However, these mistakes often stem from the automatic enactment of established patterns, executed without engaging in active cognitive evaluation to assess the suitability of the actions in the given context.

The following flowchart further expounds on this topic.

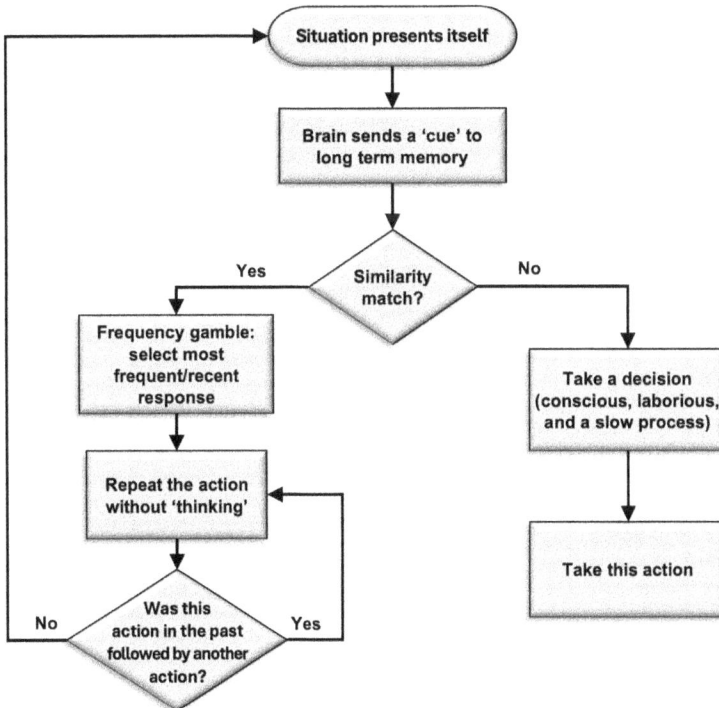

Figure 13. Automaticity.

"Similarity matching" refers to the process of deductive reasoning, 'Have I seen a similar pattern before?' – and the more similar it is, the more likely the pattern will be followed, even when it later turns out to not be a good match. "Frequency gambling" on the other hand is how often – and recently – a pattern has been repeated in the past.

Limitations of Cognitive Thinking in Altering Patterns

As discussed, cognitive thinking does not come into play when a person is engaged in an automatic behaviour; they simply follow the well-trodden behaviour pattern (refer to the flowchart above).

Traditional managerial tools, rooted in linear causality principles, often prove ineffective in transforming these patterns. Culture and cultural behaviours deviate from linearity, with groups sometimes engaging in non-objective, immeasurable or unproductive activities.

Rather than reacting linearly, attempting to address behaviour patterns with cognitive change methods sometimes causes the group culture to disproportionately over- or under-react to these change interventions. This is true for both major and minor changes imposed on the group.

The reason for this is that human behaviour is a complex adaptive system. Complex adaptive systems are dynamic networks of interacting agents – in this case, the group members – that can adapt and evolve independently from an attempt to centrally direct the change. The adaptation is complex, non-linear and unpredictable. It also most often happens with very little cognitive thinking involved, as discussed before. To put it bluntly, changes in the behaviour patterns/culture happen – or do not happen – irrespective of what the change agent intended.

Not recognising this complexity, organisations attempting to measure culture and cultural change come up hard against the inherent complexity of something not easily quantifiable or representable in graphs. Often, organisations resort to using surveys to gather individuals' opinions on values and other cultural aspects. However, the sum of individual perspectives does not necessarily equate to the collective view of the group, and this easily leads to significant discrepancies.

Additionally, these surveys must simplify complex and often subconscious cultural elements into abstract questions. For instance, a question like, "Do you feel your supervisor takes safety seriously?" requires respondents to crystallise thoughts on concepts they might previously not have consciously considered. This can lead to skewed responses, influenced by recent interactions rather than a true reflection of consistent attitudes. Therefore, the process of measuring culture through such methods can be fraught with oversimplifications and inaccuracies.

Culture Change Approaches and Challenges

Cognitive Decision-Making

When people are required to make deliberative and specific decisions, cognitive thinking is very appropriate; each choice is evaluated on its merits. This involves a process of thoughtful consideration, reasoning, and weighing potential outcomes.

Errors can occur, but they typically follow a logical thinking process that allows for the identification and analysis of where the reasoning may have faltered. In turn, this offers the flexibility to revise decisions and the decision-making approach itself. This may involve acquiring additional knowledge, seeking more data, engaging in deeper reflection, consulting with others, and reevaluating initial assumptions. While all these adjustments require considerable effort and resources, they enable a commitment to make the most informed and optimal decisions possible.

Behaviour Patterns

Transforming behavioural patterns on the other hand is a far more complex process, as these patterns are typically ingrained and function beyond the scope of logical, cognitive decision-making. Simply put, thinking alone is not enough to change these patterns. Since patterns are formed through repetition, altering them also requires consistent and repeated efforts. It follows that simply advising someone to "not do this or that" is often ineffective.

Moreover, trying to induce change through motivational strategies like reward and punishment systems may not be very effective. These approaches are grounded in conscious, cognitive activities, which do not directly address the root of habitual behaviours.

This does not imply that the change process is beyond influence. Rather, it involves guiding the target group to adopt new behavioural practices, focusing on actions rather than solely relying on cognitive persuasion techniques.

The use of the term 'influence' in this context is deliberate. Altering a group's culture is akin to social engineering, a process typically met with resistance. Therefore, it is crucial to ensure the process is transparent and subtle, fostering an environment where individuals do not feel trapped or coerced. People should always have the freedom to respond, whether in agreement or resistance, without the pressure of rewards or penalties.

Practical Implications and Guidelines

As a reminder, Chapter 11 identified the creation of a positive safety culture as one of the contributions the organisation can make to assist workers when they operate independently in the safety envelope. In this chapter, the clear message is that culture is a very complex construct, and it cannot fully be controlled by outsiders. While this is true, it does not mean that change agents cannot do anything. The following provides some practical guidelines for planning a culture intervention, followed by more detailed actions planning how to implant some key safety behaviour patterns in a work group's culture.

Practical guidelines

The following are practical guidelines to keep in mind when a change intervention is planned:

Focus on group dynamics: Change is more effectively implemented at the group level. Most individuals tend to adapt to these group dynamics, and when the changes start taking hold, the group may subtly enforce conformity among its members.

Break down changes into manageable parts: Aim to divide the desired changes into the smallest feasible components. Targeting specific elements of a pattern is often more effective than trying to overhaul an entire pattern at once.

Simplify and do not over-explain: When the changes are sufficiently small, detailed explanations may be unnecessary. Explaining engages cognitive thinking, which is not effective in modifying habitual behaviours. Managers are often told that communication is the key. Here, as long as the changes are small and at the behavioural level, doing more is more important than talking more.

Emphasise positive actions: Focus on what should happen, rather than what should not. It is easier to adopt new, positive actions than to cease negative ones. It is much more difficult to practise not doing something.

Repetition is key: Encourage the repetition of desired actions. Use various methods to prompt repetition, but avoid the misconception that you need to persuade or convince – repetition itself is powerful.

Allow time for change: Avoid pressuring the group, as this can lead to resistance and resentment. If resistance is encountered, it may indicate that the steps for change were too ambitious or that there was undue pressure to change.

Subtly create opportunities for change: Facilitate chances for the group to engage in the new behaviours without explicitly highlighting these as 'changes'. The notion of 'change' can provoke resistance due to its abstract and cognitive nature.

Influence through early adopters: Over time, early adopters will begin to incorporate the new behaviour, and if they are influential members of the group, they will be guiding the entire group's transition through their example.

Expect Imperfections: Recognise that patterns are complex and adaptive. Groups will interpret and adjust changes in unique ways, with potential over-reactions or under-reactions being a normal part of the process.

Avoid reward or punishment systems: These are cognitive strategies and are generally ineffective in changing habitual behaviours, except when the rewards or punishments are extreme, which is not desirable.

Address unwanted behaviours constructively: Instead of directly combating undesirable behaviours, refocus on practising desired actions until they become predominant within the group's patterns, thereby starving the past unwanted behaviours.

Refrain from quantitative measurements: Cultural changes are qualitative and not readily quantifiable. Attempting to measure such changes numerically can be counterproductive, i.e., counting is a cognitive activity.

Remember: Cultural change is an ongoing process requiring continuous reinforcement, due to the process known as entropy; that is, the inevitable gradual decline of a complex system. Think of it like an athlete's practice; even the most skilled athletes never cease to hone their skills.

Safety Management Application

As stated before, it is important to remember where behavioural change fits into the bigger picture: it plays a part in the safety envelope where the activities are not determined by outsiders but by the operators themselves. As discussed in Chapter 10, while the operators have a free choice of what they do inside this envelope, they are supported by the change-management actions of the organisation.

This specifically refers to including positive safety behaviours in the workers' culture, without dictating to them what they must or must not do. The following is one practical way of achieving this:

Identify safety risks and desired behaviour: Start by listing around 10 key risks that workers are likely to encounter operating within the safety envelope. For each risk, determine a few desired behaviours. Brainstorm this by engaging in a mental exercise where the change agents imagine observing a group of workers who handle these risks flawlessly. The observers should articulate what they see these workers doing correctly, focusing specifically on behaviour rather than abstract concepts.

Select and highlight exemplary behaviour: The team should then sift through this brainstormed list and pick out the top two to four behaviours for each risk that are most commendable or 'medal-worthy'. This process is repeated for all identified risks, creating a comprehensive list of exemplary behaviour.

Prioritise key behaviour to focus on: With all the nominated behaviours compiled, rank them, starting from the most desirable, and select the top ones for attention. It is important not to overextend; a focused list of perhaps 10 to 15 behaviours, with a maximum of 20, is ideal. This allows for concentrated practice without dilution. Think of a professional golfer who works on improving specific aspects of their game through repetitive practice rather than overhauling their entire technique at once.

Promote desired behaviour: Create various innovative opportunities to reinforce these behaviours. This could involve a range of methods, from visual aids like posters and call-cards to more interactive approaches like games and challenges. Also, use text and email reminders, sports bibs, or even collectible cards that depict the ideal behaviour. The goal is to expose workers to the behaviour in a subtle, non-directive way.

Focus on one behaviour at a time: Dedicate a week or two to emphasise each selected behaviour, saturating the workspace with the behaviour. Be careful, adoption or not is not the prerogative of the change agents, it belongs to the workers. Initially, there may be some scepticism and other cultural blockages, and acceptance may only grow as the intervention progresses. This is all to be expected; this is how a complex adaptive system operates.

Progressively shift focus: After extensively promoting one behaviour, gradually shift the focus to the next, while still acknowledging the previously emphasised behaviour when appropriate. This helps in maintaining a continuous, evolving focus on behaviour improvement.

Ongoing process: Recognise that this is a continuous journey. The aim is to perpetually reinforce and promote desirable behaviour, ingraining them into the team's automatic mind culture.

It is also important to note that resistance to change often correlates with the scale of the change. Slight resistance to less critical behaviour might be encountered, and in such cases, it may be prudent to temporarily deprioritise these behaviours. However, remember that a complex adaptive system changes with each adjustment, indicating that incremental changes can lead to significant overall improvements.

Finally, while focusing on these behaviours, it is essential not to worry about the risks associated with them; the focus should now be solely on the best behaviour we want to see in the team. Additionally, do not be concerned about good behaviour that did not make the final list; employees can often generalise specific behaviour to similar situations not explicitly addressed in the change process. Remember the 'similarity matching' process in Figure 13.

Remember the following key points:

Avoid negative focus: Always concentrate on creating positive safety behaviours, not on correcting the negative ones.

No rewards or punishments: Simply acknowledge the desired behaviour when observed. Avoid rewards or punishments, even verbal praise. A simple wink of acknowledgement is sufficient – we do not want a cognitive discussion about it unless the workers initiate the discussion. And even then, the approach is behaviourally focused.

Do not audit the process: Recognise that behaviour change is a complex and adaptive process. It may not follow a predictable pattern, and different individuals may change at different rates and due to different triggers.

The change is subtle: The change intervention is not a 'project'. It should not have a 'project name' and progress should not be reported in traditional management reports. It is a creative new way of life, and the process is never completed.

Avoid relying on cognitive thinking: Cognitive thinking often fails in changing behaviours that operate automatically. Behaviours that are a result of repeated actions or patterns bypass cognitive processes.

Inefficacy of traditional management tools: Traditional tools are based on linear causality, assuming a straightforward cause–effect relationship. However, cultural behaviours do not always fit this mould. Therefore, techniques based solely on cognitive activities like telling, training or threatening the operator are often ineffective.

Human behaviour is very unpredictable and does not follow straight lines: Unless extreme reward or punishment is applied, cultural/behavioural change is driven from within a group and not so much by external forces.

Misconception about performance measurement: The idea that you can measure and quantify every aspect of cultural change is a fallacy. Often, attempts to measure such changes can even be counterproductive.

This approach ensures a sustainable and positive change in workplace behaviour.

Chapter 15

Mental Wellness

The mental health of workers is a vital part of health and safety. First, mentally healthy employees are more productive, engaged and capable of making sound decisions, directly impacting workplace safety and efficiency. Even more important is the reality that organisations employ the whole person and cannot step away from the mental harm workers may suffer at work even though this is less obvious. It is a moral imperative, but also a legal and ethical responsibility of employers, and Safety 2.1 applies as much to the mental health of workers as it applies to physical safety.

Creating mental wellness in an organisation is not an event, nor is it mainly about creating awareness that mental health is important. It is creating systems and processes that will create a workplace where people can work and feel mentally safe.

One of the important factors to consider is the overlap between what happens in the workplace and outside the work environment. While businesses can be good corporate citizens and help to make society as a whole a better place, the primary focus is on the workplace.

Safety 2.1 is focused on how to make the workplace mentally safe. For this reason, it is important to understand that the focus is not to make unhealthy people healthy; the programme is primarily aimed at making sure the workplace is not mentally harming people. This is in line with the general purpose of workplace health and safety – preventing harm in the workplace.

Mental Wellness at Work Model

The model outlined below details 10 factors influencing workplace wellness. Improving these areas is not just beneficial for mental wellness but also constitutes sound business practice. This includes aspects like leadership styles, reward systems, workload management, and nurturing interpersonal relations, all sound business practices in their own right. Yet, it is crucial to revisit these elements with a focus on mental safety.

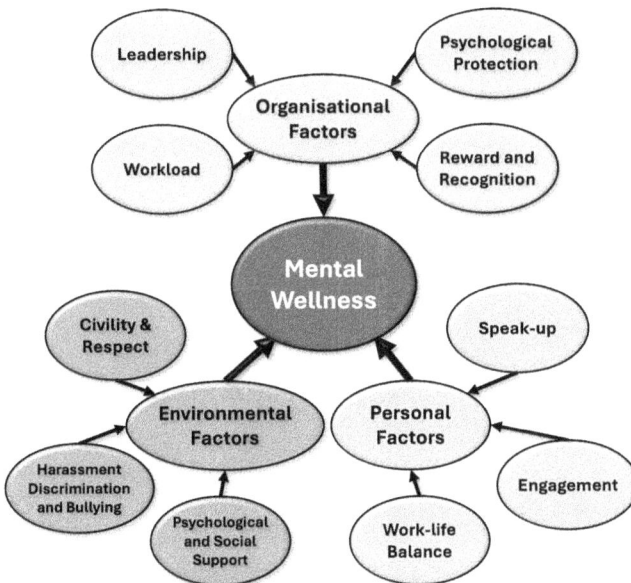

Figure 14. Mental Health Model.

A. Organisational Factors

1. Leadership and expectations

The influence of leadership style on mental well-being is substantial. Leaders who demonstrate commitment to long-term objectives and embed a deep sense of mission, vision and purpose not only function as catalysts for change but also substantially mitigate workplace stress. Their ability to engage, motivate and inspire employees is crucial in cultivating a positive and productive work environment. Aspects like communication, involving employees in decision-making, and aligning organisational goals with employee welfare all contribute to mental well-being. Additionally, leaders should model healthy work-life balance practices, encourage open dialogue, and show genuine concern for employees' well-being, thereby fostering a supportive and inclusive culture.

2. Psychological protection

Creating a psychologically safe workplace where employees feel empowered to contribute ideas, raise concerns and express disagreements without fear of negative repercussions is essential. Such openness not only drives innovation and effective problem-solving but also significantly bolsters mental well-being. This environment is characterised by regular employee check-ins, availability of mental health resources, and a culture of empathy and understanding. Leaders in such environments are approachable and receptive, encouraging employees to share their thoughts and feelings, and ensuring that these contributions are valued and considered.

3. Recognition and reward

Regular, culturally sensitive recognition of employees' performance and achievements is essential for enhancing morale. Celebrations, recognitions and acknowledgments should respect individual preferences for private or public recognition, addressing diverse needs and sensitivities. Personalising rewards to align with individual achievements and cultural backgrounds enhances the sense of value among employees. Leaders should also ensure that recognition is timely, relevant and aligned with the company's values, thereby reinforcing desired behaviours and outcomes.

It is important to distinguish this from the application of recognition and reward as tools for facilitating cultural change. Within this context, rewarding and recognising staff is not a strategy to drive specific changes; it is simply a way to express gratitude for their hard work and dedication.

4. Workload management

It is vital to ensure that employees have manageable workloads and access to necessary resources like time, equipment and support. This involves not simply controlling the volume of work but also improving the quality of work conditions. Regular workload assessments, flexible work arrangements, and a culture that addresses overtime and overwork proactively are crucial. After the impact of Covid on working practices, remote working options have become more common and should be considered responsibly.

Leaders should also focus on providing clear job descriptions, setting realistic deadlines and encouraging regular breaks to prevent burnout.

B. Environmental Factors

5. Social support

Fostering a culture of trust, support and emotional integration among colleagues and supervisors is fundamental. This includes providing help when needed and cultivating a perception of strong organisational support. Creating peer support networks, mentorship programmes and team-building activities can enhance the sense of trust and emotional safety. This approach promotes a sense of community within the workplace, allowing for more effective collaboration and communication.

6. Harassment, discrimination and bullying

Implementing a zero-tolerance policy towards harassment, discrimination and bullying is non-negotiable. Proactive measures to protect all employees and promote inclusivity and equal opportunities are essential. This policy should be supported by clear reporting mechanisms, prompt and equitable disciplinary actions, and continuous education and training to foster a respectful work environment.

It is essential to acknowledge that the act of reporting itself poses risks for targeted individuals. Therefore, feedback mechanisms should be designed with sensitivity to this inherent risk.

7. Civility and respect

Maintaining a respectful and considerate workplace is vital. This includes dignified interactions among employees, clients and the public. Regular training and reinforcement of values like dignity, good manners, cultural sensitivity and the avoidance of offensive behaviours help in creating an environment where respect and civility are integral. Leaders should model these behaviours and address any breaches promptly and effectively.

C. Personal Factors

8. Speaking up

Encouraging open communication and emotional expression, including in traditionally male-dominated environments, is crucial. All employees should feel safe to voice their concerns and experiences without fear of judgement or reprisal. This includes establishing regular feedback sessions, anonymous reporting options, and fostering a culture where every voice is valued. Leaders should actively seek out and respond to employee feedback, ensuring that concerns are addressed and acted upon.

9. Engagement and involvement

Fostering an environment where employees feel connected, motivated and passionate about their work is key to mental wellness. This can be achieved by involving employees in goal-setting, in decision-making processes, and providing opportunities for professional growth. When employees find their work meaningful and feel it contributes to a greater good, their job satisfaction and mental health improve. Leaders should focus on aligning individual roles with the company's mission and values, thereby enhancing engagement and purpose.

10. Work-life balance

Recognising and supporting a healthy balance between work and personal life is critical. This involves acknowledging potential conflicts between work and personal responsibilities and offering solutions such as flexible working hours and remote work options. It is important to create a culture where employees can discuss their work-life balance needs openly and without fear. Leaders should also be role models in maintaining their work-life balance, for example, demanding "work phones off at 6 pm", thereby setting a positive example for their teams.

Implementation

Methods of implementing the Mental Wellness Model will vary, based on the existing organisational culture. In a 'toxic' culture, direct intervention from senior leadership might be necessary to remove initial barriers. In contrast, a more positive culture might benefit from collaborative forums involving managers and employees to tackle each factor.

It is important not to rely solely on statistics or culture surveys to gauge success, as they might not capture the nuances of an environment and can mask underlying issues. Biometric data may reveal gender-based disparities but might not capture the experiences of gender minorities. Anecdotal evidence, while less specific, can offer rich insights in a healthy environment and indicate a reluctance to share in an unhealthy one.

Another common pitfall to avoid is over-reliance on cognitive interventions, like online training programmes. These often fail to change behaviours, as merely informing people about support channels does not necessarily empower them to use them, especially in environments where expressing vulnerability is stigmatised. A more effective approach involves analysing each factor and devising specific improvement plans. These might include policy changes, work method modifications, external support access, feedback mechanisms and, yes, sometimes training. The implementation of these measures might vary in complexity and timescale and could range from directive actions (like banning gender-based jokes) to fostering gradual cultural shifts (like encouraging social support).

A typical improvement methodology might include:

Project Facilitation: Assign a project facilitator who possesses not only expertise in research but also skills in group facilitation. This should be a senior person; the individual should have a track record of successfully leading diverse teams and managing complex projects. Their role will be pivotal in guiding the group through various stages of the project, ensuring that every member's voice is heard and valued.

Working Group: Convene a dynamic discussion forum, composed of a balanced mix of managers and influential workers. This group should be led by a senior leader who acts as a sponsor, providing not just resources but also lending credibility to the initiative. The senior leader's involvement signals the organisation's commitment to the project's objectives. The primary aim of this forum is to meticulously identify areas within the model that demand attention. For this group to operate effectively, it is crucial to cultivate a high-trust group culture. Trust within this context is not merely a feeling but a cultural phenomenon – as explored in detail in Chapter 14, which focuses on culture change. It is essential to employ the methods outlined in that chapter to facilitate the development of trust within the group.

Initial Assessment: In the initial phase, the forum conducts a thorough assessment of various factors within the organisation and selects one or more as focal points. This stage involves deep analysis and understanding of the underlying issues that need addressing. The selection process should be democratic, allowing every group member to voice their opinions and concerns.

Strategy Development Process: Here, the facilitator plays a critical role in steering the group towards discovering and researching potential strategies for improvement. This phase is not just about identifying strategies but also involves evaluating their feasibility, potential impact and alignment with the organisation's overall goals. The facilitator should encourage creative thinking and foster an environment where innovative ideas can be freely shared and discussed.

Implementation and Monitoring Procedures: Once strategies are chosen, the next step involves their meticulous implementation. This stage requires careful planning and resource allocation. Monitoring the progress of these strategies is equally important. Regular updates should be communicated to the entire organisation, celebrating successes and acknowledging individual and team contributions. This transparency helps maintain momentum and ensures everyone stays informed and engaged.

Iterative Process Description: The forum's work is ongoing and iterative. After the implementation of initial strategies, the forum should regroup to identify new factors that require attention. This continuous cycle ensures that the organisation is always moving forward, adapting and improving. The process should be dynamic, with the flexibility to incorporate new insights and learnings from previous phases.

Measurement of Wellness

A repeated refrain in this book is that a complex adaptive system cannot be measured by using traditional methods. Surveys and hard data on graphs will never succeed in measuring the state of mental wellness in an organisation.

While it is not a good idea to attempt measuring mental wellness at all, if the organisation insists on some form of measurement, neutral observers, ideally from outside the organisation, may be used to conduct focus group discussions and then report on their findings. This is called 'phenomenological research' and if appropriate, safety professionals are encouraged to further research this construct.

Part 4

Summary and Concluding Thoughts

Chapter 16
Then the Penny Drops

The transition to Safety 2.1 is a complex and nuanced process that is not immune to misinterpretation. At first glance, it might appear that the principles of Safety 2.1 could easily integrate within the conventional Safety 1 framework. One might even anticipate that over time, the identification of potential hazards and the corresponding increase in the quantity and specificity of controls could proliferate. The concept of a safety envelope, while theoretically appealing, may in practice prove to be as constraining as ever.

This highlights a crucial reality: Safety 2.1 delves much deeper than mere risk management; it embodies a holistic approach that encompasses every aspect of safety management.

Consider for example incident investigations, a key component of the continuous improvement drive in safety management. Under Safety 1, the primary focus of incident investigations is learning from past errors to prevent recurrence. This involves a thorough analysis to identify root causes and contributing factors, followed by the development of strategies to mitigate these causes in the future.

While the intention behind this approach is commendable, it often results in an ever-expanding set of prescriptive measures for safe work practices. This trend, highlighted by Dekker,[*] can lead to the bureaucratisation of safety, invariably stifling flexibility and innovation.

The Swiss Cheese Model developed and popularised by James Reason,[†] exemplifies the traditional approach to incident investigation. By analysing the sequence of events and identifying the failures that allowed an incident to occur, this model aims to fortify defences against future failures. Yet, this method epitomises Safety 1's focus on pinpointing and rectifying errors, assuming that all causes can be found and fixed. This is typical cause-and-error thinking at the core of Safety 1.

Transitioning to Safety 2.1 introduces a paradigm shift in incident investigation methodologies. Unlike its predecessor, Safety 2.1 emphasises recognising and reinforcing what goes right, in addition to analysing what went wrong.

In the realm of Safety 2.1, investigations into incidents significantly diverge from the traditional model. Rather than dissecting the incident to identify causal factors, the investigation examines whether the controls at the boundary of the safety envelope functioned as intended. These controls are designed to mitigate risk to a level the organisation can tolerate. The dynamics within the safety envelope, governed by the operators and the complex adaptive systems they navigate, remain beyond the scope of formal investigation.

---

[*] Dekker, S. W. (2014). The bureaucratisation of safety. *Safety Science* 70: 348–357.
[†] Reason, J. (1997). *Managing the risks of organisational accidents*. Ashgate.

This is a crucial point and worth repeating: an investigation after an event ends at the edge of the safety envelope. There are no inquiries into why the operator did or did not perform a specific action. The investigation aims only to ensure that the measures intended to protect operators were effective. It may verify if a guard was in place against a spinning machine or if a critical procedure was followed, provided these measures were identified to reduce the likelihood of harm to within the agreed risk tolerance level.

Moreover, concerns that learnings from incidents might not be disseminated reflect Safety 1 thinking. This perspective underestimates the capability of complex adaptive systems to communicate and enforce necessary adjustments among operators, who do not intend harm in their actions.

The Penny Drops

At this moment, the realisation hits: we are witnessing not just a rebranding of old ideas but a significant departure from traditional safety-management approaches. The shift to Safety 2, and its evolution into Safety 2.1, profoundly affects how safety is managed. Leadership practices that assign blame or mandate adherence to an ever-growing number of protocols in response to incidents are misaligned with the ethos of Safety 2.1. Such actions, often rooted in a blame culture, starkly contrast with the principles of Safety 2.1. Safety 2.1 ensures controls are in place to reduce the risk to the edge of the safety envelope, and thereafter trust the people to do the right things. It recognises the ability of the peer group to self-regulate and does not assume correction primarily sits outside the complex adaptive system's self-organising abilities.

Consider another element of typical safety-management systems: performance management. Safety 1 is fixated on quantifying metrics, such as the lost time injury frequency rate (LTIFR) or its counterpart, the total recordable injury frequency rate (TRIFR). Even the focus on 'lead indicators', counting activities to justify improvement, like the number of hazards or near misses reported, falls into this category.

Unfortunately, within a complex adaptive organisation, individual managers may resort to 'creative accounting' to enhance their safety statistics. For instance, injury statistics might be manipulated by attributing incidents to occurrences outside of work, attempting to reduce the severity of injuries from lost-time to medical-treatment-only injuries, or by blaming workers for exploiting the system. Simultaneously, they may emphasise reporting minor 'positive events', such as hazards reported, to bolster their 'lead indicators'.

This means that safety practitioners may spend significant effort attempting to increase the quality of statistics reported, despite the questionable value of many of these events.

In contrast, Safety 2.1 recognises that safety performance is not easily quantifiable; it cannot be solely counted. It is a phenomenon, observable in reality, but often difficult to measure quantitatively. Its value lies in the quality of events, many of which are subtle and even obscured.

'You know it when you see it', but it is challenging to quantify. Reporting, therefore, uses different lenses to provide insight into the status of safety performance, recognising and acknowledging that this reporting focuses on iterative improvement and stakeholder involvement. These lenses include ongoing monitoring of risk controls, bringing risk down to the edge of the safety envelope, as well as functions supporting operators inside the envelope: providing a safe working environment and adequate tools, supporting their learning and skills development, positively influencing cultural behaviour patterns, and supporting their mental well-being.

Extending Beyond Safety Management

A study of the dynamics of Safety 2.1 shows that these truths extend far beyond the confines of safety management, touching upon broader organisational behaviours. Safety performance, as a subset of organisational behaviour, is no different from any other organisational behaviour, including productivity enhancement, scheduling, resource management and others. The same processes described here can equally be applied to these organisational spheres.

This poses very specific challenges to safety professionals: how they can become true change agents in both safety management practices and the interconnectedness of all organisational behaviours. This challenge is not trivial; it will require confronting deep-seated thinking, with many traditional leaders overtly or covertly resisting change and hanging on traditional cause-and-effect approaches. However, by adopting the safety envelope concept, i.e. a safe place where the people can take responsibility for their own behaviour, organisations can pioneer a management style that transcends traditional reward and punishment schemes, grounded in the widely discredited principles of behaviourism. This approach advocates for a modern management philosophy, informed by robust theoretical underpinnings, which recognises and leverages the expertise and adaptability of frontline operators.

This narrative established in this discussion aims not only to contribute to theoretical discourse but, more importantly, to facilitate the practical application of established theories. It targets safety practitioners and organisational leaders willing to embrace change, emphasising that effective safety management, while crucial, is part of a broader organisational context.

In its application, this approach will assist safety practitioners in becoming true professionals, influencing changes well beyond the narrow confines of merely being safety technicians.

Bibliography

Anderson, D., & Anderson, L. A. (2010). *How command and control as a change leadership style causes transformational change efforts to fail.* Being First, Inc.

Balducci, C., Baillien, E., Broeck, A. V. D., Toderi, S., Fraccaroli, F. (2020). Job demand, job control, and impaired mental health in the experience of workplace bullying behavior: A two-wave study. *International Journal of Environmental Research and Public Health* 17(4): 1358.

Bargh, J. A. (1994). The four horsemen of automaticity: Awareness, intention, efficiency, and control in social cognition. In R. S. Wyer Jr. & T. K. Srull (Eds.), *Handbook of social cognition: Basic processes; Applications* (pp. 1–40). Lawrence Erlbaum Associates, Inc.

Blewett, V., & O'Keeffe, V. (2011). Weighing the pig never made it heavier: Auditing OHS, social auditing as verification of process in Australia. *Safety Science* 49(7): 1014–1021.

Borys, D. J., Else, D., & Leggett, S. (2009). The fifth age of safety: the adaptive age. *Journal of Health and Safety Research and Practice* 1(1): 19–27.

Canadian Mental Health Commission. (2013). *Psychological health and safety in the workplace: Prevention, promotion, and guidance to staged implementation.* First ed. Bureau de normalisation du Quebec.

Canadian Standards Association. (2013). *Psychological health and safety in the workplace.* Canadian Standards Association.

Conklin, T. (2019). *Pre-accident investigations: An introduction to organisational safety.* CRC Press.

Dekker, S. (2014). *Safety differently: Human factors for a new era.* CRC Press.

Dekker, S. (2018). *I Am not a Policy Wonk.* Blog via www. safetydifferently.com

Dekker, S. W. (2014). The bureaucratisation of safety. *Safety Science* 70: 348–357.

Dekker, S., Cilliers, P., & Hofmeyr, J. H. (2011). The complexity of failure: Implications of complexity theory for safety investigations. *Safety Science* 49(6): 939–945.

Dibley, L., Dickerson, S., Duffy, M., & Vandermause, R. (2020). *Doing hermeneutic phenomenological research: A Practical Guide.* Sage.

DNM Safety. *Systems thinking for safety: Ten principles. A white paper.* EUROCONTROL.

Engel, D., Woolley, A. W., Jing, L. X., Chabris, C. F., & Malone, T. W. (2014). Reading the mind in the eyes or reading between the lines? Theory of mind predicts collective intelligence equally well online and face-to-face. *PLOS ONE* 9(12): e115212.

Gardner, B., Rebar, A. L., & Lally, P. (2019). A matter of habit: Recognising the multiple roles of habit in health behaviour. *British Journal of Health Psychology* 24(2): 241–249.

Groenewald, T. (2004). A phenomenological research design illustrated. *International Journal of Qualitative Methods* 3(1): 42–55.

Hale, A., & Borys, D. (2013). Working to rule or working safely? Part 1: A state of the art review. *Safety Science* 55: 207–231.

Hale, A., & Borys, D. (2013). Working to rule or working safely? Part 2: The management of safety rules and procedures. *Safety Science* 55: 222–231.

Harkema, S. (2003). A complex adaptive perspective on learning within innovation projects. *The Learning Organisation* 10(6): 340–346.

Hollnagel, E. (2013). A tale of two safeties. *Nuclear Safety and Simulation* 4(1): 1–9.

Hollnagel, E., Wears, R. L., & Braithwaite, J. (2015). *From Safety-I to Safety-II: a white paper. The resilient health care net*: published simultaneously by the University of Southern Denmark, University of Florida, USA, and Macquarie University, Australia.

LaMontagne, A. D., Milner, A. J., Allisey, A. F., Page, K. M., Reavley, N. J., Martin, A., Tchernitskaia, I., Noblet, A. J., Purnell, L. J., Witt, K., & Keegel, T. G. (2016). An integrated workplace mental health intervention in a policing context: Protocol for a cluster randomised control trial. *BMC Psychiatry* 16(1): 1–13.

Lee, N. K., Roche, A., Duraisingam, V., Fischer, J. A., & Cameron, J. (2014). Effective interventions for mental health in male-dominated workplaces. *Mental Health Review Journal* 19(4): 237–250.

Lowell, K. R. (2016). An application of complexity theory for guiding organisational change. *The Psychologist-Manager Journal* 19(3–4): 148.

MacQueen, J. (2020). *The flow of organisational culture*. Palgrave Macmillan.

Matud, M. P., Lópes-Curbelo, M., & Fortes, D. (2019). Gender and psychological well-being. *International Journal of Environmental Research and Public Health* 16(19): 3531.

McDougall, M., Ronkainen, N., Richardson, D., Littlewood, M., & Nesti, M. (2020) Three team and organisational culture myths and their consequences for sport psychology research and practice. *International Review of Sport and Exercise Psychology* 13(1): 147–162.

Milner, A., Kavanagh, A., King, T., & Currier, D. (2018). The influence of masculine norms and occupational factors on mental health: Evidence from the baseline of the Australian longitudinal study on male health. *American Journal of Men's Health* 12(4): 696–705.

Nguyen, N. C., & Bosch, O. J. (2014). The art of interconnected thinking: Starting with the young. *Challenges* 5(2): 239–259.

Pascale, R. T. (1999). Surfing the edge of chaos. *Sloan Management Review* 40(3): 83–94.

Pickering, A., & Cowley, S. P. (2010). Risk Matrices: Implied accuracy and false assumptions. *Journal of Health and Safety Research and Practice* 2: 11–18.

Raafat, R. M., Chater, N., Frith, C. (2009). Herding in humans. *Trends in Cognitive Sciences* 13(10): 420–428.

Rathje, S. (2009). The definition of culture: An application-oriented overhaul. *Interculture Journal* p. 35.

Reason, J. (2017). *The human contribution: Unsafe acts, accidents and heroic recoveries*. CRC Press.

Reason, J. (1997). *Managing the risks of organisational accidents*. Ashgate.

Reyneke, P. (2011). *A complex adaptive systems approach to reducing work related injuries in New Zealand supermarkets*. Doctoral dissertation, CQUniversity.

Rubin, M., Paolini, S., Subašić, E., & Giacomini, A. (2019). A confirmatory study of the relations between workplace sexism, sense of belonging, mental health, and job satisfaction among women in male-dominated industries. *Journal of Applied Social Psychology* 49(5): 267–282.

Sherratt, F., & Ivory, C. (2019). Managing "a little bit unsafe": Complexity, construction safety and situational self-organising. *Engineering, Construction and Architectural Management*, 26(11): 2519–2534.

Stalinski, S. (2004). Organisational intelligence: A systems perspective. *Organisation Development Journal* [serial online]. 22: 55–67.

Sweeney, A. M. (2016). *The making of a habit: The moderating role of construal level on the development of automaticity* (Doctoral dissertation, State University of New York at Stony Brook).

Templeton, K., Bernstein, C. A., Sukhera, J., Nora, L. M., Newman, C., Burstin, H., Guille, C., Lynn, L., Schwarze, M. L., Sen, S., & Busis, N. (2019). Gender-based differences in burnout: Issues faced by women physicians. *NAM Perspectives*. Discussion Paper, National Academy of Medicine, Washington, DC. May 30, 2019.

Woolley, A. W., & Malone, T. W. (2011). *Defining collective intelligence: Phenomenon, possibility, or paradox?* In A. Kosbelt (Ed.), *The Cambridge Handbook of Creativity*. Cambridge University Press.

Woolley, A. W., Chabris, C. F., Pentland, A., Hashmi, N., & Malone, T. W. (2010). Evidence for a collective intelligence factor in the performance of human groups. *Science* 330(6004): 686–688.

About the Author

Paul Reyneke – pronounced "Ray-nuh-kuh" – is a distinguished health and safety professional with a robust academic background. He holds a Professional Doctorate (DProf) in organisational behaviour from CQUniversity and a Master of Management (MMgt) from the University of Auckland. Paul has received numerous accolades, including the prestigious New Zealand Safety Practitioner of the Year award in 2007 (recipients only receive the award once) and induction into the Beta Gamma Sigma honorary society at the UOA Business School, recognising his outstanding academic achievements.

Throughout his career, Paul has held senior health and safety roles in major organisations, leading teams across New Zealand,

Australia, Thailand, China, and the USA. Known for his slightly unconventional approach, Paul firmly believes that individuals, when equipped with the necessary skills and tools, are best at managing their own safety and the safety of those around them.

This book encapsulates Paul's legacy in the field. As he passes the baton to the next generation of safety professionals, he envisions a future where the profession evolves, leveraging the collective power of people rather than increasing bureaucracy.

Contact Paul: paul@adapto.co.nz
www.adapto.co.nz

www.ingramcontent.com/pod-product-compliance
Lightning Source LLC
Chambersburg PA
CBHW040902210326
41597CB00029B/4938